Human and Mosquito Lysozymes

Mauro Prato
Editor

Human and Mosquito Lysozymes

Old Molecules for New Approaches Against Malaria

Editor
Mauro Prato
Neuroscience
University of Torino
Torino, Italy

ISBN 978-3-319-38089-6 ISBN 978-3-319-09432-8 (eBook)
DOI 10.1007/978-3-319-09432-8
Springer Cham Heidelberg New York Dordrecht London

Printed on acid-free paper

Springer is part of Springer Science+Business Media (www.springer.com)

To Wall-E
Beloved, smart, and inspiring rabbit

Preface

The important thing in science
is not so much to obtain new facts
as to discover new ways
of thinking about them.

—Sir William Henry Bragg,
1915 Nobel prize in Physics

In *The Wonderful Wizard of Oz*, Dorothy Gale heads for the Emerald City by following the yellow brick road, in order to get back home. It is no secret that eventually she will return to Kansas only thanks to the silver shoes she has worn since the beginning of her journey. As a child, I enjoyed reading Dorothy's adventures, and dreamed of getting to know the Scarecrow and the Cowardly Lion (though I have never been a big fan of the Tinman, no offense intended). As an adult, I am now aware that from moral imperatives to philosophical reflections to political plots, L. Frank Baum offers a magic box of profound discoveries, buried in a playground of childhood whimsy. Indeed, the hallmark of superb writing lies in the ability to compress multiple layers of meaning into a single narrative.

On a certain level, science is not much different, as it resembles a complex yet exciting book that is just waiting to be unrolled by those avid readers commonly known as "researchers." Biology makes no exception, and such a perspective fits well with the experimental evidence discussed in the present work. Addressing either students or specialists, the authors will introduce them to an innovative research field, which is ironically based on the consolidated knowledge of an almost 100-year-old molecule.

Now, let's take a step back. Despite the considerable early successes achieved by the global eradication program launched a few years ago (2007) by charity foundations such as the Bill & Melinda Gates Foundation and Rolling Back Malaria, also endorsed by World Health Organization, malaria remains an alarming emergency in developing countries. Therefore, any parasite or host molecules serving as new affordable markers for early diagnosis of disease complications or as new targets for vector control need to be urgently identified. Intriguingly, old well-known enzymes such as lysozymes have been proposed in recent years as very relevant molecules to be targeted either in mosquito or in human hosts, possibly paving the way for the future development of innovative and cost-effective tools to fight malaria.

Lysozymes are antibacterial proteins widely present in the animal kingdom and defined by their ability to hydrolyze β-1,4-glycosidic linkage between *N*-acetylmuramic acid and *N*-acetylglucosamine of peptidoglycan in the cell wall of bacteria. Recently, a few independent research groups have reported some interesting evidence on the involvement of both human and mosquito lysozymes in malaria. Hemozoin, a lipid-bound ferriprotoporphyrin IX crystal produced by *Plasmodium* parasites after hemoglobin catabolism, was shown to promote in vitro the early release of human lysozyme from adherent monocytes. Consistently, the plasma levels of human lysozyme and the number of hemozoin-containing leukocytes in the peripheral blood of *P. falciparum*-infected patients correlated well with parasitemia degree and disease severity. On the other hand, the mosquito homologue of human lysozyme was shown to facilitate the development of *Plasmodium berghei* and *falciparum* within *Anopheles gambiae*, *stephensi*, and *dirus* vectors after binding to oocysts. All these data will be thoroughly discussed in this book in an attempt to show the readers the potential role of these molecules in malaria diagnosis and vector control approaches.

My editing of *Human and Mosquito Lysozymes: Old Molecules for New Approaches Against Malaria* has been rather time consuming, yet very rewarding in providing me with a privileged look at the status of cutting edge research in this field. I would like to thank all of the authors (Nicoletta Basilico, Sarah D'Alessandro, Luca Facchinelli, Giuliana Giribaldi, Clelia Oliva, Manuela Polimeni, Roberta Spaccapelo, and Vivian Tullio) for their hard work in preparing excellent discussions on their respective topics. Additionally, I am very grateful to Springer's staff members (Rita Beck, Susan Safren, and Arthur Smilios) for inviting me on board and helping me during such an exciting journey. And, of course, thanks to you, who are reading this book. Will lysozymes reveal themselves as the "silver standard" to leave the Land of Malaria? Let's knock our heels together three times and command the shoes to…

Torino, Italy Mauro Prato

Contents

Chapter 1
Etiopathogenesis and Pathophysiology of Malaria

Giuliana Giribaldi, Sarah D'Alessandro, Mauro Prato, and Nicoletta Basilico

1 Introduction

According to the latest estimates, World Health Organization (WHO) officially registered a decline in malaria mortality rates by about 42 % globally and by 49 % in the WHO African Region between 2000 and 2012. During the same period, malaria incidence rates declined by 25 % around the world and by 31 % in the African Region. However, this progress, a likely consequence of the global eradication program launched by charity foundations (Khadjavi et al. 2010), is no cause for complacency. The absolute numbers of malaria cases and deaths are not going down as fast as they could. The disease still took an estimated 627,000 lives in 2012, mostly those of children under 5 years of age in Africa among more than 200 million clinical cases worldwide, of which 91 % were due to *Plasmodium falciparum* (WHO 2013). Malaria remains an alarming emergency in developing countries: an estimated 3.4 billion people were at risk of malaria in 2012. Of this total, 2.2 billion were at low risk (<1 reported case per 1,000 population), of whom 94 %

G. Giribaldi (✉)
Dipartimento di Oncologia, Università degli Studi di Torino, Torino, Italy
e-mail: giuliana.giribaldi@unito.it

S. D'Alessandro
Dipartimento di Scienze Farmacologiche e Biomolecolari, Università degli Studi di Milano, Milan, Italy

M. Prato
Dipartimento di Neuroscienze, Università degli Studi di Torino, Torino, Italy

Dipartimento di Scienze della Sanità Pubblica e Pediatriche, Università degli Studi di Torino, Torino, Italy

N. Basilico
Dipartimento di Scienze Biomediche, Chirurgiche e Odontoiatriche, Università degli Studi di Milano, Milan, Italy

M. Prato (ed.), *Human and Mosquito Lysozymes*,
DOI 10.1007/978-3-319-09432-8_1

were living in geographic regions other than the African Region. The 1.2 billion at high risk (>1 case per 1,000 population) were living mostly in the African Region (47 %) and the Southeast Asia Region (37 %) (WHO 2013). In 2007, the Bill and Melinda Gates Foundation, rapidly endorsed by the WHO and the Roll Back Malaria association, claimed for malaria eradication as the primary goal to be prosecuted (Roberts and Enserink 2007).

Malaria is a parasitic disease confined mostly to the tropical areas, caused by parasites of the genus *Plasmodium* and transmitted by mosquitoes of the genus *Anopheles*. Annually, nearly a 2.37 billion people are estimated to be at risk of infection by *P. falciparum*, the most virulent among *Plasmodia* (Khadjavi et al. 2010). Cyclical fever has always been considered as a typical symptom of malaria, long before the identification of *Plasmodium parasites* as the etiological agents of the disease. As reported by Oakley and colleagues (Oakley et al. 2011): "…early records of fever that can be attributed to malaria can be found in Bronze Age texts from China, India and Greece that include the writings of Huang Ti, the Atharvaveda, and early Greek medical works. In these accounts, there are references to symptoms such as tertian and quartan fevers, enlarged spleen, and the association of the disease with monsoon weather. Definitive accounts of malaria occur in classical Indic (Charaka and Sushuruta) and Greek (Hippocrates) medical texts. These texts not only document the classification of malaria fevers but also the association of these fevers with residence in marshy places."

Responsible for human infections are typically four species of *Plasmodium*, *P. malariae*, *P. ovale*, *P. vivax*, *P. falciparum* (Tilley et al. 2011), that can affect humans in more than 90 countries, inhabited by 40 % of the global population. In some of these areas, over 70 % of residents are continuously infected by the most deadly form of the parasite, *P. falciparum*, able to provoke the majority of death and the most severe clinical manifestation (Rowe et al. 2009). Surviving children develop various levels of natural immunity; however, it does not protect them from repeated infections and illness throughout life. In 2008, a fifth species, *P. knowlesi*, has been associated to infection in human in Southeast Asia with *P. falciparum*, *vivax*, and *malariae* (Cox-Singh et al. 2008). In Northern Africa malaria is not particularly widespread and it is caused by *P. vivax*. In Eastern and Central Africa predominant is *P. falciparum* but *P. vivax* and *P. malariae* are also present while in Western Africa predominant is *P. falciparum* but also present are *P. ovale* and *P. vivax*. In India predominant is *P. vivax* but it is possible to find also *P. falciparum* while in Central America and in island of Pacific Ocean there are *P. vivax* and *P. falciparum* (Tan et al. 2014).

The malaria parasite was observed for the first time in the blood cells of people with malaria in 1880 by Alphonse Laveran, at Costantina in Algeria, who also demonstrated the link between the disease and the protozoa, winning the Nobel Prize for medicine in 1907 (see Biography of Alphonse Laveran, The Nobel Foundation). In 1886, in Pavia (Italy), Camillo Golgi (Nobel Prize in 1906) was the first one associating malaria fever with the cyclical release of malaria parasites during schizont rupture of red blood cells (RBCs) and demonstrating that *P. vivax* was responsible for benign tertian fever (every 48 h) and *P. malariae* for quartan fever (every 72 h) (Golgi 1886). Subsequently Ettore Marchiafava, Angelo Celli, Amico Bignami, and

Giuseppe Bastianelli, in Rome, recognized the existence of *P. falciparum*, responsible for malignant tertian fever. In 1894, Patrick Manson, in China, was the first one to hypothesize that the *Plasmodium* vector was a mosquito. This thesis was demonstrated in 1897 by Ronald Ross (Nobel Prize 1902), in India (see Biography of Ronald Ross, The Nobel Prize Foundation). In 1898, Anopheles species of mosquitoes were identified in Rome as the vector of the disease by Giovanni Battista Grassi, who described also the parasite cycle in the different species of *Plasmodia* (Grassi 1900).

2 *Plasmodium falciparum* Life Cycle

As schematized in Fig. 1.1, *P. falciparum* life cycle involves stages either in mosquito vector (sexual reproduction) or in human host (asexual replication), where in particular the "blood stage" is responsible for much of the disease pathology. Parasites are transmitted to humans by the females of the *Anopheles* mosquito species. There are about 460 species of *Anopheles* mosquitoes, but only 68 transmit

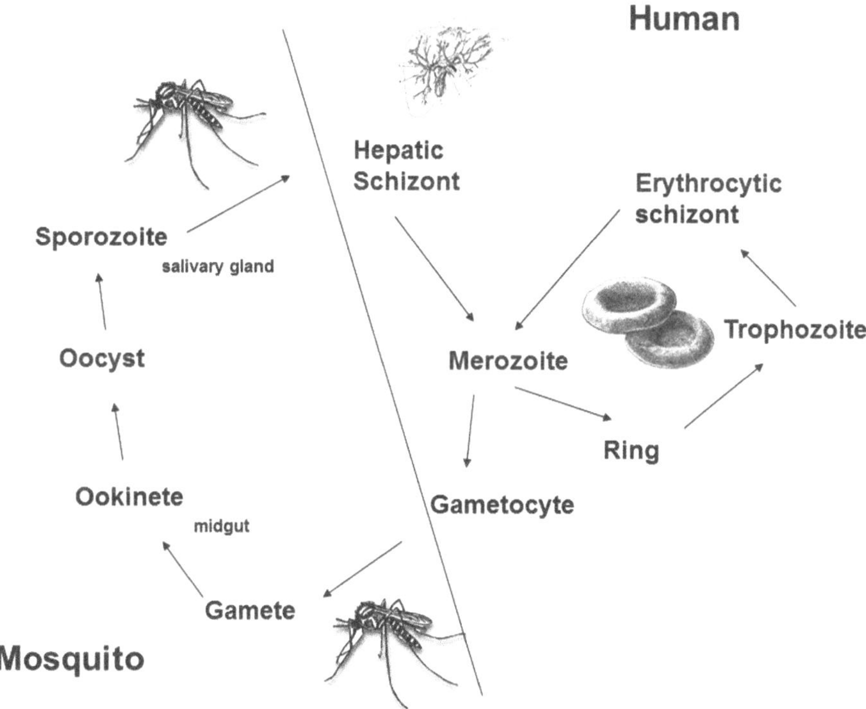

Fig. 1.1 *Plasmodium falciparum* life cycle. *P. falciparum* life cycle involves either stages in mosquito vector (sexual reproduction) or in human host (asexual replication). The "blood stage" is responsible for much of the disease pathology in humans (see also Fig. 1.2)

malaria. *Anopheles gambiae*, found in Africa, is one of the major malaria vectors. It is long living, prefers feeding on humans, and lives in areas near human habitation (Rogier and Hommel 2011).

The intensity of malaria parasite transmission varies geographically according to vector species of *Anopheles* mosquitoes. Risk is measured in terms of exposure to infective mosquitoes, with the heaviest annual transmission intensity ranging from 200 to >1,000 infective bites per person. Interruption of transmission is technically difficult in many parts of the world because of limitations in approaches and tools for malaria control.

The malarial infection begins when the sporozoite stage of the parasite that resides within the salivary gland of the mosquito halts in the host liver (Ménard 2005). This happens when an infected female bites a healthy person and takes its blood meal, injecting a small amount of saliva into the skin wound. Male mosquito does not feed on blood, hence only female serves as a vector. The saliva contains antihemostatic and anti-inflammatory enzymes disrupting the clotting process and inhibiting the pain reaction. Typically, each infected bite contains 5–200 sporozoites which proceed to infect the human vector. Sporozoites stay for prolonged time in the skin before reaching the blood stream. Only the parasites surviving to phagocytes can rapidly enter the human bloodstream through a blood vessel, where they circulate for a matter of minutes before infecting liver cells (Ejigiri and Sinnis 2009).

2.1 Liver Stage in Man

After circulating in the bloodstream, the sporozoites migrate to the liver and finally infect a hepatocyte, after crossing several Kupffer cells and hepatocytes (Mota et al. 2001). The sporozoites rapidly grow in size absorbing nourishment to form a large round schizont. The schizont divides by schizogony, a type of asexual reproduction, in which multiple fissions result in the formation of a number of small, spindle-shaped uninucleate cells called merozoites (Rogier and Hommel 2011). After schizonts rupture, merozoites are released into the sinusoids or venous passages of the liver. This phase of asexual reproduction is called preerythrocytic schizogony. The merozoites are immune to medicines and host natural resistance. After a development stage in liver, during which there are no clinical symptoms of the disease, merozoites are released into the blood and enter the erythrocytic portion of their life cycle. A single schizont can produce thousands of merozoites by asexual reproduction.

In *P. falciparum* and *P. malariae*, schizont development and rupture occur rapidly, and merozoites can begin the erythrocyte invasion 1 or 2 weeks after the invasion of hepatocytes. This phase of the cycle is different in *P. ovale* and in *P. vivax* where parasites can stay in the liver as dormant cells (hypnozoites) for months or years before returning to schizont form and relapse the initial infection (Markus 2011).

The invasion of erythrocytes by merozoites involves some proteins originating from both parasites and RBCs. In particular, major merozoite surface proteins (MSP)-1 and -9 bind to erythrocyte band 3 protein (Kariuki et al. 2005), the merozoite orients its apical end towards the erythrocyte surface through merozoite apical membrane antigen-1 transmembrane protein (Mitchell et al. 2004) and penetrates in the RBC involving erythrocyte binding antigens and the *P. falciparum* reticulocyte-binding homologs, which binds glycophorins and other unknown receptors (Baum et al. 2005). In *P. vivax*, reticulocyte invasion occurs after interaction with the Duffy blood group antigen, the erythrocyte receptor for the chemokine Interleukin-8/CXCL8. This antigen is generally missing in African population that is why *P. vivax* results more common in tropical areas other than Africa, such as Southern Asia and Malaysia (Horuk et al. 1993). By invading the erythrocytes, the merozoites initiate the blood stage of the asexual cycle. The blood stage takes around 48 h to be completed. During this period, the parasites develop through several stages (ring, trophozoite, and schizont forms, respectively), characterized by different structures and specialized stage-specific features (Bannister et al. 2000).

2.2 *Erythrocytic Stage in Man*

The merozoites feed on erythrocytes, become rounded and gradually modify into trophozoites. During growth, a vacuole appears in the center of merozoites and the nucleus is pushed to one side; this modification, that is known as "ring stage," gives it a ring-like appearance in Giemsa-stained blood smears. The parasite by its cytostome feeds on hemoglobin and other nutrient taken from the plasma. The food vacuole secretes some digestive enzymes, which break down hemoglobin into proteins and hematin. Proteins are used by the parasite as nourishment source, whereas hematin is converted into a waste product called hemozoin (Hz).

As the ring form of the parasite enlarges, it begins to synthesize some stage-specific molecules, which can be exported into the RBC (Das 1994), modifying the RBC membrane which now begins to adhere to the linings of visceral and other blood vessels, including those of the placenta (Pouvelle et al. 2000). The parasite eventually grows into the more rounded trophozoite form. During this stage most active feeding, growth, and RBC modifications occur. New molecules are exported into the RBC: some of them assemble into flat membranous sacs of various forms, including Maurer's clefts, which are visible in stained smears (Aikawa et al. 1986; Atkinson and Aikawa 1990). Other proteins interact with the RBC membrane and cytoskeleton to form small knobs (Atkinson and Aikawa 1990; Cooke 2000) on its surface. Some molecules such as *P. falciparum* erythrocyte membrane protein (*Pf*EMP)1 allow the infected RBC to adhere to the vascular endothelium, thus counteracting the action of immune defenses aimed at removing parasites from the blood stream through the spleen. As a consequence of infected RBC cytoadherence to brain–blood vessel walls, some malaria complications may occur, including cerebral malaria (Adams 2002) and placental malaria (Scherf et al. 2001). Other exported

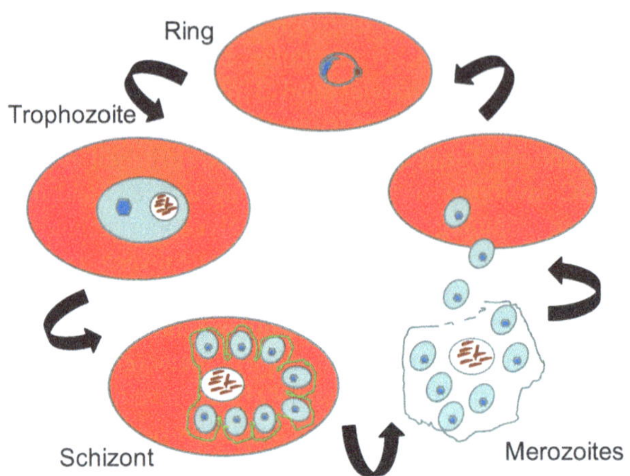

Fig. 1.2 *Plasmodium falciparum* erythrocytic cycle. The merozoites infect the erythrocyte and modify into "ring stage." The ring form of the parasite enlarges and grows into the trophozoite form. The parasite converts hemoglobin into dark crystal particles to form the malarial pigment (Hz). The trophozoites metamorphose into schizonts that divide to form daughter nuclei of new merozoite cells in the erythrocyte. Finally the merozoites burst from the RBC, proceeding to infect other erythrocytes

molecules increase RBC permeability to nutrients. The parasite continues feeding on hemoglobin, and the heme products derived from hemoglobin digestion are converted into dark crystal particles to form the malarial pigment (Hz), which is scattered within the digestive vacuole (Bannister and Mitchell 2003).

During their growth, the trophozoites metamorphose into schizonts (Rogier and Hommel 2011). Schizonts appear after a period of about 36–40 h of growth and represent the full-grown trophozoites. Schizonts carry out some nuclear divisions and an intense synthesis and assembly of molecules to organize RBC invasion. The nucleus of the schizont divides in the following 6–8 h to form from 12 to 24 daughter nuclei of new merozoite cells in the erythrocyte. This phase of asexual multiplication is known as erythrocytic schizogony. Finally the RBC membrane and parasitophorous vacuolar membrane are lysed through a protease-dependent process (Salmon et al. 2001) and the merozoites burst from the RBC, proceeding to infect other erythrocytes.

Free merozoites are very small (1.2 mm of length). Still, they contain all the tools deemed necessary to invade new RBCs as detailed by Bannister and Mitchell (Bannister and Mitchell 2003). A single erythrocytic cycle is completed in 48 h (see Fig. 1.2). Parasite remain in the bloodstream for roughly 60 s before entering into another erythrocyte, restarting the process (Cowman and Crabb 2006).

The infection cycle occurs in a highly synchronous fashion, with roughly all of the parasites throughout the blood at the same stage of development. The toxins are liberated into the blood along with the liberation of merozoites and then deposed into the liver and the spleen or under the skin, so that the host gets a sallow color.

The accumulated toxins cause malaria fever: the patient suffers from chills, shivering, sweating, high temperature headache, abdominal and back pain, nausea, diarrhea, and sometimes vomiting. The fever lasts for 6–10 h and then it comes again after every 48 h with the liberation of a new generation of merozoites. During the erythrocytic stage, some merozoites increase in size to form two types of gametocytes, the macrogametocytes and microgametocytes. The macrogametocytes (female) are large, round with the food-laden cytoplasm and a small eccentric nucleus. The microgametocytes (male) are small, with clear cytoplasm and a large central nucleus. This process is called gametocytogenesis. The specific factors underlying this sexual differentiation are largely unknown. The gametocytes take roughly 8–10 days to reach full maturity and do not develop further until they get sucked by the appropriate species of mosquito. If this does not happen, they degenerate and die, since they require lower temperature for further development (Prato et al. 2011).

2.3 Life Cycle in Mosquito

When a female *Anopheles* sucks the blood of a malaria patient, the gametocytes enter along with blood, reaching the stomach and leading to formation of gametes (Aly et al. 2009). Only the gametocytes survive inside the stomach, while the other forms of the parasites, as well as the erythrocytes, are digested. Two types of gametes are formed: microgametocyte (male) and the macrogametocyte (female). The microgametocytes become active and their nucleus divides to produce 6–8 haploid daughter nuclei. The nuclei arrange at the periphery. The cytoplasm gives out the same number of flagella-like projections. A daughter nucleus enters in each projection. These projections separate from the cytoplasm. This process of formation of microgametes is called exflagellation. From each microgametocyte, 6–8 flagella-like active microgametes are formed. The megagametocyte undergoes some reorganization and forms the megagametes. Fertilization of the female gamete by the male gamete occurs rapidly after gametogenesis. The fertilization event produces a zygote that remains inactive for some time and then elongates into a worm-like ookinete or vermicule. The zygote and ookinete are the only diploid stages. The ookinete penetrates the wall of the stomach and comes to lie below its outer epithelial layer. It gets enclosed in a cyst formed partly by the zygote and partly by the stomach of mosquito. The encysted zygote is called oocyst. The oocysts absorb nourishment and grow to about five times in size. They protrude from the surface of the stomach as transparent rounded structures. Over a period of 1–3 weeks, the oocyst grows to a size of tens to hundreds of micrometers. During this time, multiple nuclear divisions occur. As a consequence of oocyst maturation, the oocyst divides to form multiple haploid sporozoites. Each oocyst may contain thousands of sporozoites and groups of sporozoites get arranged around the vacuoles. This phase of asexual multiplication is known as sporogony. In the mosquito, the whole sexual cycle is completed in 10–21 days. Finally the oocyst bursts and sporozoites are liberated into the hemolymph of the mosquito. They spread throughout the hemolymph and eventually reach the salivary glands and enter the duct of the hypopharynx. The mosquito now becomes infective and the sporozoites get inoculated or injected into the human blood when the mosquito bites, starting a new life cycle. It is estimated that a single infected mosquito may contain as many as 200,000 sporozoites (Prato et al. 2011).

3 Hemozoin (Malarial Pigment)

During the intraerythrocytic stages (especially the trophozoite and schizont stages) of malaria infection, the parasites degrade the hemoglobin present in the cytoplasm of the RBC. This process occurs mostly inside the food vacuole of the parasite that in *P. falciparum* is an acidic structure with an estimated pH of 5.0–5.4 (Krogstad et al. 1985; Klonis et al. 2007). The digestive vacuole lacks the characteristic lysosomal phosphatases and glycosidases present in the vacuoles of other organisms and is dedicated almost exclusively to hemoglobin degradation. Such an event was believed to play a vital role for malaria parasites, being an essential source of nutrients and energy (Francis et al. 1997), as they had been observed to have a limited capacity to synthesize them on their own (Coronado et al. 2014). However, this process generates heme, which is highly toxic to the parasites. The parasite is unable to excrete free toxic heme and does not possess any heme oxygenases, a class of enzymes responsible for the degradation of the heme moiety. Therefore, *Plasmodium* aggregates the heme monomer into an insoluble inert biocrystal called malarial pigment or Hz (Slater 1992). In particular, during hemoglobin digestion, alpha-hematin (ferriprotoporphyrin IX), which is toxic to the parasite, is released (Orjih 2001) and—most possibly as a protection strategy for the parasite—transformed into Hz (chemically identical to beta-hematin), a molecule with paramagnetic properties.

The structure of Hz has been elucidated by X-ray diffraction, infrared spectroscopy, Raman spectroscopy, and chemical synthesis. The molecule was first proposed to consist of an unusual polymer of heme groups linked by bonds between the oxygen from the carboxylate of one heme and the central ferric ion of the next heme. This unusual linkage allows the heme units to be aggregated into an ordered insoluble crystal, representing a unique way of heme processing (Slater et al. 1991). It is now well known that Hz is composed of hematin molecules bonded by reciprocal iron–carboxylate bonds to the propionic side chains of each porphyrin to form dimers, which are further connected via hydrogen bonds to form a triclinic crystal (Pagola et al. 2000). This was concluded through photoacoustic spectroscopy (Balasubramanian et al. 1984) and corroborated more recently by using beta-hematin DMSO solvate, being this the first Fe(III) PPIX model of Hz studied by single crystal X-ray diffraction (Gildenhuys et al. 2013).

In the past, Hz was only considered to be a metabolic waste. However, this molecule has been shown to sequester in various organs including lung, liver, spleen, and brain, indicating that Hz can potentially contribute to the development of malaria immunopathogenesis (Francis et al. 1997; Deroost et al. 2013). It has been suggested that Hz could be related to the complications in severe malaria (see next paragraph) (Lalloo et al. 2007). Following the rupture of *Plasmodium*-infected RBCs, Hz is released from the parasite digestive vacuole and is rapidly engulfed by phagocytes (Olivier et al. 2014).

As a result of phagocytosis, several functions of monocytes are seriously compromised, including repeated phagocytosis and oxidative burst (Schwarzer et al. 1992), bacterial killing abilities (Fiori et al. 1993), MHC Class II expression and antigen presentation (Scorza et al. 1999), maturation to dendritic cells (Urban and Todryk 2006), and coordination of erythropoiesis (Giribaldi et al. 2004) (see also Chap. 5 for further details). However, impaired Hz-laden human monocytes do not undergo apoptosis: apparently, Hz-dependent enhanced expression of anti-apoptotic HSP-27 leads to prolonged monocyte survival, thereby contributing to maintain the impaired monocytes in the bloodstream (Giribaldi et al. 2010; Prato et al. 2010a). Moreover, Hz-fed monocytes show enhanced gene expression of a large number of proinflammatory molecules, including cytokines (TNFalpha, IL-1beta, IL-1RA) and chemokines (MIP-1alpha/CCL-3, MIP-1beta/CCL-4, GROalpha/CXCL-1, GRObeta/CXCL-2, GROgamma/CXCL-3, MCP-1/CCL-2, IL-8/CXCL-8, ENA-78/CXCL-5) (Giribaldi et al. 2010). Hz also upregulates the expression and activity of few monocytic enzymes. As more extensively reviewed previously (Prato and Giribaldi 2011), Hz was shown in a series of works to increase RNA/protein expression, protein release, and proteolytic activity of matrix metalloproteinase-9 (MMP-9), a cytokine-related proteolytic enzyme that has been proposed to participate in a mechanism that could explain the remarkable release of cytokines in malaria infection. Indeed, cytokines such as TNFalpha and IL-1beta are able to induce the expression and activity of MMP-9 in culture of monocytes after phagocytosis of Hz (Prato et al. 2005, 2008). Higher levels of MMP-9 were also found in brain of cerebral malaria-sensitive *P. berghei*-infected mice and it was shown by immunohistochemistry that the monocytic cells were responsible for this production (Van den Steen et al. 2006). Since MMP-9 gene transcription is induced by TNFalpha and active MMP-9 can shed membrane-bound TNFalpha into the extracellular environment through proteolytic cleavage, a pathological loop involving TNFalpha and MMP-9 is established, leading to abnormal levels of cytokine production (Prato et al. 2005). Data from in vitro cultures of human monocytes showed that some peroxidated lipids attached to Hz may play a crucial role either in enhanced MMP-9 expression and activity or in TNFalpha and IL-1beta production (Prato et al. 2008, 2010b). On the other side, it has been shown that lipid-free synthetic pigment beta-hematin is able to perform cleavage of the proform of MMP-9 in vitro suggesting that heme core may concur in its activation (Geurts et al. 2008). As such, the interaction between Hz and MMP-9 appears a promising research field for new therapies in complicated malaria, also taking in account that drugs against these enzymes are already available because of their role in other diseases, including cancer and neuroinflammation.

4 Malaria: Severe Complications

P. falciparum malaria ranges from asymptomatic infections to the classic symptoms of malaria (e.g., fever, chills, sweating, headache, and muscle aches), which in a subpopulation of cases result in severe life-threatening complications such as cerebral malaria (CM), severe malarial anemia (SMA), respiratory distress (RD), and acute renal failure (ARF) (Perkins et al. 2011). Although the pathophysiology of malaria is multifactorial and only partially understood, there are some important interactions between the parasite and the host that can determine the clinical outcome of the disease: endemicity patterns, acquisition of naturally acquired malarial immunity, parasite virulence, multiplication rate, antigenic variation, polymorphic variability in either human host or *Plasmodium* parasite, and age (Newton and Krishna 1998; Abdalla and Pasvol 2004). It is well known that children typically display enhanced susceptibility to SMA, while nonresident malaria naïve adults progress towards ARF and RD due to pulmonary edema (White 1998).

4.1 Cerebral Malaria

A paradigmatic complication of *falciparum* malaria is CM, which develops after infected RBCs sequester in the microvasculature of the central nervous system (CNS). *P. falciparum* is unique as it causes mature infected RBC to sequester and adhere to microvascular beds in numerous organs. Unlike the other human malarial parasites which rarely cause neurological dysfunction, *P. falciparum*-induced CM often leads to death or severe neurological sequelae (Grau and Craig 2012). However, *P. falciparum* appears to remain in the vascular space without ever entering the brain parenchyma, thus raising question of how intravascular *Plasmodium* parasites are capable of inducing a devastating neural dysfunction in CM. Recent evidence suggests that a compromised integrity of the blood–brain barrier (BBB) results in a subsequent increase in BBB permeability which enables toxic soluble factors released by either host or parasite to cross this barrier and exert neurological effects (Coltel et al 2004).

According to the WHO clinical criteria, CM is defined as a potentially reversible, diffuse encephalopathy causing a Glasgow coma score of 11/15 or less, often associated with fitting, in the absence of other factors that could cause unconsciousness such as coexistent hypoglycemia or other CNS infections (Medana and Turner 2006). CM appears as a diffuse encephalopathy commonly presenting with headache, agitation, frank psychosis, seizures and impaired consciousness, and occasionally with brainstem signs or focal neurological signs such as hemiplegia and cranial nerve palsies (Newton and Warrel 1998; Gitau and Newton 2005). It is difficult to confirm diagnoses of CM in endemic areas because of overlapping infections such as bacterial meningitis in patients showing incidental malarial parasitemia (Berkley et al. 1999). Children from areas endemic for malaria or nonimmune adults traveling from developed countries are at higher risk for developing CM. On the contrary, CM is rarely encountered in >10-year-old patients who have been exposed to *P. falciparum* since birth. Mortality ranges from 15 to 30 %, and 11 % of children display neurological deficits upon discharge (Newton and Warrel 1998).

The pathophysiological mechanisms underlying CM are not fully understood so far. There are currently three distinct theories on the etiology of CM typical features: (1) the mechanical hypothesis; (2) the permeability hypothesis; and (3) the humoral hypothesis (Polimeni and Prato 2014). The mechanical hypothesis proposes that CM is caused by a mechanical obstruction of the cerebral microvasculature by infected RBC, with coma resulting from impaired brain perfusion (Pino et al. 2005). The permeability hypothesis suggests that a leaky BBB allows toxic compounds to enter the brain and cause neurological dysfunction (Gitau and Newton 2005). Such a hypothesis has been natural extended to the humoral hypothesis, conceiving that host factors such as leukocyte-derived cytokines and chemokines can enter the brain parenchyma after reduced BBB integrity, thus causing CM symptoms such as fever and coma (Pino et al. 2005; Gitau and Newton 2005).

In the last decade, experimental evidence implicated MMPs in malaria pathogenesis (Khadjavi et al. 2010; Szklarczyk et al. 2007; Prato et al. 2011; Geurts et al. 2012; Piña-Vázquez et al. 2012; Polimeni and Prato 2014). MMPs are host proteolytic enzymes involved in degradation of basement membranes, disruption of inter-endothelial tight junctions, and cleavage of a large spectrum of proinflammatory, membrane-bound and hemostasis-related molecules, and they may play a crucial role in CM. Further in-depth analysis of the involvement of MMPs in CM might help to design new adjuvant therapies. It is well understood that MMP inhibitors could prevent BBB leakage and reduce the exacerbated inflammatory response, thus reducing the high mortality rates of CM patients, along with the frequency of neurological sequelae in recovering patients.

4.2 Severe Malarial Anemia

In regions with high rates of malaria and human immunodeficiency virus (HIV), including sub-Saharan Africa, the majority of infants and young children suffer from anemia (Lartey 2008). As shown by Brabin and colleagues (Brabin et al. 2001) SMA is an important childhood health problem also if numerous efforts aimed at reducing the burden of anemic patients. SMA consists of values of hemoglobin concentration lower than 5.0 g/dL or hematocrit lower than 15.0 % (WHO 2000).

Lysis of infected and uninfected RBC (Dondorp et al. 1999; Price et al. 2001), dyserythropoiesis and bone marrow suppression (Phillips et al. 1986), splenic sequestration of RBCs (Buffet et al. 2009), coinfections with hookworm bacteremia and HIV-1 (Berkley et al. 2005; Otieno et al. 2006; Bassat et al. 2009; Davenport et al. 2010; Were et al. 2011), and chronic transmission of malaria in holoendemic regions characterize the SMA etiology. It should be noted that these factors can lead to chronically low hemoglobin values observed in infants and young children living in holoendemic regions (Perkins et al. 2011).

Plasmodium-triggered hemolysis participates in reducing hemoglobin levels in pediatric patients. However, impaired and/or ineffective erythropoiesis represents the primary mechanism underlying low hemoglobin levels in children with SMA.

This translates into failure in the ability to replenish the reduced pool of erythrocytes due to parasite- and/or antimalarial-driven hemolysis. Studies of Dormer and colleagues (Dörmer et al. 1983) provide some quantitative data on the extent of "parenchymal damage" of bone marrow and stress the impact of ineffective erythropoiesis and reduced rate of erythropoietic proliferation on the emergence of anemia in patients with acute *falciparum* malaria. The results of a large scale ex vivo study in anemic children attending a rural hospital in Mozambique suggest that hemozoin in the bone marrow has a role in the pathogenesis of malarial anemia through ineffective erythropoiesis (Aguilar et al. 2014).

Perkins and coworkers affirm that reduced erythropoiesis in SMA children results from imbalanced inflammation (see Perkins et al. 2011 for a more exhaustive review). Indeed, to tentatively control the parasitemia, the host releases a large array of pro- and anti-inflammatory cytokines, chemokines, growth factors, and effector molecules as part of the innate immune response. However, the immune response to malaria can result in successful control of the parasitemia only depending on the magnitude and timing of the release of inflammatory mediators. If the inflammatory milieu is not properly balanced, it can seriously damage the host, also suppressing the erythropoietic response. They also show as some parasitic products drive the innate immune response in the infected human host, including Hz, glycosylphosphatidylinositols (GPIs), and parasitic antigens. Phagocytosis of Hz by circulating monocytes, neutrophils, and resident macrophages represents a primary factor for stimulating the innate immune response to *P. falciparum*. Among other mechanisms, Hz generates an innate immune response through the toll-like receptor (TLR) system.

4.3 Respiratory Distress

Several types of respiratory distress can occur in malaria patients. Hyperventilation is a consequence of metabolic acidosis. This is mainly due to the accumulation of lactate, produced by the parasite or by peripheral tissues in hypoxic conditions. Metabolic acidosis is often worsened by renal dysfunction. As a consequence of lower blood pH, respiratory rates are increased to expel more carbon dioxide, resulting in hyperventilation. Therefore, hyperventilation should be considered a peripheral or renal pathology rather than pulmonary (Van den Steen et al. 2013). Whether hyperventilation may contribute to pulmonary edema—for instance by prolonged high tidal volume breathing—is currently topic of debate, and the simultaneous occurrence of both events cannot be excluded (Nayak et al. 2011). Hyperventilation represents the main respiratory distress occurring in African children with malaria and is also frequent in adults (Marsh et al. 1995; English et al. 1996).

Other types of respiratory distress are acute lung injury (ALI) and acute respiratory distress syndrome (ARDS). ARDS is a severe condition associated with high mortality rates. It is characterized by marked, diffuse alveolar inflammation, damage to the alveolar–capillary membrane, alveolar edema, and severe hypoxemia. Common causes of ARDS include sepsis, bacterial or viral infectious diseases, and aspiration pneumonia (Matthay and Zemans 2011; Wheeler and Bernard 2007), all factors that are often concomitant in malaria patients. However, ARDS may also present as a severe complication of malaria independently of any other cause (Taylor et al. 2006, 2012; Mohan et al. 2008). Malaria-associated (MA)-ARDS occurs mainly in adults and lethality rates reach 80 % of patients, despite antimalarial treatment. The largest part of ARDS cases is registered in adult patients in low-transmission areas, as well as in nonimmune travelers. On the contrary, resident adults from high-transmission areas appear semi-immune and protected against severe malaria. Pregnant women with placental malaria are also at higher risk to develop MA-ARDS (Taylor et al. 2006, 2012; Mohan et al. 2008). Depending on the study area and the parasite species, the estimated prevalence rates range from 2 to 20 % in adults with severe malaria (see Mohan et al. 2008 for a more detailed review). Several case reports have been published for each human malaria parasite species. MA-ARDS appears to be the most prevalent complication of *Plasmodium knowlesi* infections (59–70 % of severe cases) (Daneshvar et al. 2009; William et al. 2011). Notably, besides antimalarial chemotherapy, positive pressure ventilation measures are currently the only available remedy. ARDS has been proposed to likely represent the most extreme form of pulmonary involvement in malaria (Maguire et al. 2005).

4.4 Acute Renal failure

ARF is more frequent in *P. falciparum* infections, although *P. vivax* and *P. malariae* can occasionally be associated with renal impairment. ARF generally occurs in nonimmune adults and older children (Eiam-Ong and Sitprija 1998; Barsoum 2000; Eiam-Ong 2003). Indeed, ARF results largely higher in patients from non-malarious regions than in the areas with high malaria transmission rates. According to one report, ARF incidence is as high as that of CM (Weber et al. 1991). The mechanism underlying malarial ARF remains unknown so far. The manifestations vary from milder forms (e.g., prerenal azotemia) to more severe forms (ischemic ARF), depending on the degree of renal hypoperfusion. Mechanical obstruction by infected erythrocytes, immune-mediated glomerular and tubular pathologies, fluid loss due to multiple mechanisms, and alterations in the renal microcirculation, should be listed among the most credited hypotheses. Restricted local blood flow in the kidneys is considered to play a major role in development of malarial ARF. Low intake or high loss of fluids as a consequence of vomiting and pyrexial sweating could be responsible for dehydration and renal ischemia. Increased fluid administration, oxygen toxicity, and yet unidentified factors may contribute to pulmonary edema, ARDS, multiorgan failure, and death. Currently, the mainstay treatment consists of appropriate antimalarial drug therapy, fluid replacement, and renal replacement therapy (Das 2008).

References

Abdalla SH, Pasvol G (2004) Malaria: A hematolgical perspective. Imperial College Press, London, UK

Adams S (2002) Breaking down the blood-brain barrier: signaling a path to cerebral malaria? Trends Parasitol 18:360–366

Aguilar R, Moraleda C, Achtman AH et al (2014) Severity of anaemia is associated with bone marrow haemozoin in children exposed to Plasmodium falciparum. Br J Haematol 164: 877–887. doi:10.1111/bjh.12716

Aikawa M, Uni Y, Andrutis AT et al (1986) Membrane-associated electron-dense material of the asexual stages of Plasmodium falciparum: evidence for movement from the intracellular parasite to the erythrocyte membrane. Am J Trop Med Hyg 35:30–36

Aly AS, Vaughan AM, Kappe SH (2009) Malaria parasite development in the mosquito and infection of the mammalian host. Annu Rev Microbiol 63:195–221

Atkinson CT, Aikawa M (1990) Ultrastructure of malaria infected erythrocytes. Blood Cells 16:351–368

Balasubramanian D, Mohan Rao C, Panijpan B (1984) The malaria parasite monitored by photoacoustic spectroscopy. Science 223:828–830

Bannister L, Mitchell G (2003) The ins, outs and roundabouts of malaria. Trends Parasitol 19:209–213

Bannister LH, Hopkins JM, Fowler RE (2000) A brief illustrated guide to the ultrastructure of Plasmodium falciparum asexual blood stages. Parasitol Today 16:427–433

Barsoum RS (2000) Malaria acute renal failure. J Am Soc Nephrol 11:2147–2154

Bassat Q, Guinovart C, Sigaúque B et al (2009) Severe malaria and concomitant bacteraemia in children admitted to a rural Mozambican hospital. Trop Med Int Health 14:1011–1019

Baum J, Maier AG, Good RT et al (2005) Invasion by P. falciparum merozoites suggests a hierarchy of molecular interactions. PLoS Pathog 1:e37

Berkley JA, Mwangi I, Mellington F et al (1999) Cerebral malaria versus bacterial meningitis in children with impaired consciousness. Q J Med 92:151–157

Berkley JA, Lowe BS, Mwangi I et al (2005) Bacteremia among children admitted to a rural hospital in Kenya. N Engl J Med 352:39–47

Biography of Alphonse Laveran. The Nobel Foundation. http://www.nobelprize.org/nobel_prizes/medicine/laureates/1907/laveran-bio.html

Biography of Ronald Ross. The Nobel Foundation. http://www.nobelprize.org/nobel_prizes/medicine/laureates/1902/ross-bio.html

Brabin BJ, Premji Z, Verhoeff F (2001) An analysis of anemia and child mortality. Nutr J 131: 636S–645S

Buffet PA, Safeukui I, Milon G et al (2009) Retention of erythrocytes in the spleen: a double-edged process in human malaria. Curr Opin Hematol 16:157–164
Coltel N, Combes V, Hunt NH et al (2004) Cerebral malaria—a neurovascular pathology with many riddles still to be solved. Curr Neurovasc Res 1:91–110
Cooke BM (2000) Falciparum malaria: sticking up, standing out and outstanding. Parasitol Today 16:416–420
Coronado LM, Nadovich CT, Spadafora C (2014) Malarial hemozoin: from target to tool. Biochim Biophys Acta 1840:2032–2041. doi:10.1016/j.bbagen.2014.02.009, pii: S0304-4165(14)00055-5
Cowman AF, Crabb BS (2006) Invasion of red blood cells by malaria parasites. Cell 124:755–766
Cox-Singh J, Davis TM, Lee KS et al (2008) Plasmodium knowlesi malaria in humans is widely distributed and potentially life threatening. Clin Infect Dis 46:165–171
Daneshvar C, Davis TM, Cox-Singh J et al (2009) Clinical and laboratory features of human Plasmodium knowlesi infection. Clin Infect Dis 49:852–860
Das A (1994) Biosynthesis, export and processing of a 45 kDa protein detected in membrane clefts of erythrocytes infected with Plasmodium falciparum. Biochem J 302:487–496
Das BS (2008) Renal failure in malaria. J Vector Borne Dis 45:83–97
Davenport GC, Ouma C, Hittner JB et al (2010) Hematological predictors of increased severe anemia in Kenyan children coinfected with Plasmodium falciparum and HIV-1. Am J Hematol 85:227–233
Deroost K, Tyberghein A, Lays N et al (2013) Hemozoin induces lung inflammation and correlates with malaria-associated acute respiratory distress syndrome. Am J Respir Cell Mol Biol 48:589–600
Dondorp AM, Angus BJ, Chotivanich K (1999) Red blood cell deformability as a predictor of anemia in severe falciparum malaria. Am J Trop Med Hyg 60:733–737
Dörmer P, Dietrich M, Kern P et al (1983) Ineffective erythropoiesis in acute human P. falciparum malaria. Blut 46:279–288
Eiam-Ong S (2003) Malaria nephropathy. Semin Nephrol 23:21–33
Eiam-Ong S, Sitprija V (1998) Falciparum malaria and the kidney: a model of inflammation. Am J Kidney Dis 32:361–375
Ejigiri I, Sinnis P (2009) Plasmodium sporozoite-host interactions from the dermis to the hepatocyte. Curr Opin Microbiol 12:401–407. doi:10.1016/j.mib.2009.06.006
English M, Waruiru C, Amukoye E et al (1996) Deep breathing in children with severe malaria: indicator of metabolic acidosis and poor outcome. Am J Trop Med Hyg 55:521–524
Fiori PL, Rappelli P, Mirkarimi SN et al (1993) Reduced microbicidal and anti-tumour activities of human monocytes after ingestion of Plasmodium falciparum-infected red blood cells. Parasite Immunol 15:647–655
Francis SE, Sullivan DJ Jr, Goldberg DE (1997) Hemoglobin metabolism in the malaria parasite Plasmodium falciparum. Annu Rev Microbiol 51:97–123. doi:10.1146/annurev.micro.51.1.97
Geurts N, Martens E, Van Aelst I et al (2008) Beta-hematin interaction with the hemopexin domain of gelatinase B/MMP-9 provokes autocatalytic processing of the propeptide, thereby priming activation by MMP-3. Biochemistry 47:2689–2699
Geurts N, Opdenakker G, Van den Steen PE (2012) Matrix metalloproteinases as therapeutic targets in protozoan parasitic infections. Pharmacol Ther 133:257–279
Gildenhuys J, LeRoex T, Egan TJ et al (2013) The single crystal X-ray structure of β-hematin DMSO solvate grown in the presence of chloroquine, a β-hematin growth-rate inhibitor. J Am Chem Soc 135:1037–1047
Giribaldi G, Ulliers D, Schwarzer E et al (2004) Hemozoin- and 4-hydroxynonenal-mediated inhibition of erythropoiesis. Possible role in malarial dyserythropoiesis and anemia. Haematologica 89:492–493
Giribaldi G, Prato M, Ulliers D et al (2010) Involvement of inflammatory chemokines in survival of human monocytes fed with malarial pigment. Infect Immun 78:4912–4921
Gitau EN, Newton CR (2005) Review article: blood–brain barrier in falciparum malaria. Trop Med Int Health 10:285–292

Golgi C (1886) Sull'infezione malarica. Arch Sci Med 10:109–134
Grassi GB (1900) Studi di uno zoologo sulla malaria. Reale Academia dei Lincei, Roma
Grau GE, Craig AG (2012) Cerebral malaria pathogenesis: revisiting parasite and host contributions. Future Microbiol 7:291–302
Horuk R, Chitnis CE, Darbonne WC et al (1993) A receptor for the malarial parasite Plasmodium vivax: the erythrocyte chemokine receptor. Science 261:1182–1184
Kariuki MM, Li X, Yamodo I et al (2005) Two Plasmodium falciparum merozoite proteins binding to erythrocyte band 3 form a direct complex. Biochem Biophys Res Commun 338:1690–1695
Khadjavi A, Giribaldi G, Prato M (2010) From control to eradication of malaria: the end of being stuck in second gear? Asian Pac J Trop Med 3:412–420
Klonis N, Tan O, Jackson K et al (2007) Evaluation of pH during cytostomal endocytosis and vacuolar catabolism of haemoglobin in Plasmodium falciparum. Biochem J 407:343–354
Krogstad DJ, Schlesinger PH, Gluzman IY (1985) Antimalarials increase vesicle pH in Plasmodium falciparum. J Cell Biol 101:2302–2309
Lalloo DG, Shingadia D, Pasvol G et al (2007) UK malaria treatment guidelines. J Infect 54:111–121
Lartey A (2008) Maternal and child nutrition in Sub-Saharan Africa: challenges and interventions. Proc Nutr Soc 67:105–108
Maguire GP, Handojo T, Pain MC et al (2005) Lung injury in uncomplicated and severe falciparum malaria: a longitudinal study in Papua, Indonesia. J Infect Dis 192:1966–1974
Markus MB (2011) Malaria: origin of the term "hypnozoite". J Hist Biol 44:781–786. doi:10.1007/s10739-010-9239-3
Marsh K, Forster D, Waruiru C et al (1995) Indicators of life-threatening malaria in African children. N Engl J Med 332:1399–1404
Matthay MA, Zemans RL (2011) The acute respiratory distress syndrome: pathogenesis and treatment. Annu Rev Pathol 6:147–163
Medana IM, Turner GD (2006) Human cerebral malaria and the blood-brain barrier. Int J Parasitol 36:555–568
Ménard R (2005) Medicine: knockout malaria vaccine? Nature 13:113–114
Mitchell GH, Thomas AW, Margos G et al (2004) Apical membrane antigen 1, a major malaria vaccine candidate, mediates the close attachment of invasive merozoites to host red blood cells. Infect Immun 72:154–158
Mohan A, Sharma SK, Bollineni S (2008) Acute lung injury and acute respiratory distress syndrome in malaria. J Vector Borne Dis 45:179–193
Mota MM, Pradel G, Vanderberg JP et al (2001) Migration of Plasmodium sporozoites through cells before infection. Science 291:141–144
Nayak KC, Mohini, Kumar S et al (2011) A study on pulmonary manifestations in patients with malaria from northwestern India (Bikaner). J Vector Borne Dis 48:219–223
Newton CR, Krishna S (1998) Severe falciparum malaria in children: current understanding of pathophysiology and supportive treatment. Pharmacol Ther 79:1–53
Newton CR, Warrel DA (1998) Neurological manifestations of falciparum malaria. Ann Neurol 43:695–702
Oakley MS, Gerald N, McCutchan TF et al (2011) Clinical and molecular aspects of malaria fever. Trends Parasitol 27:442–449
Olivier M, Van Den Ham K, Shio MT et al (2014) Malarial pigment hemozoin and the innate inflammatory response. Front Immunol 5:25, eCollection 2014
Orjih AU (2001) On the mechanism of hemozoin production in malaria parasites: activated erythrocyte membranes promote beta-hematin synthesis. Exp Biol Med 226:746–752
Otieno RO, Ouma C, Onġecha JM et al (2006) Increased severe anemia in HIV-1-exposed and HIV-1-positive infants and children during acute malaria. AIDS 20:275–280
Pagola S, Stephens PW, Bohle DS et al (2000) The structure of malaria pigment beta-haematin. Nature 404:307–310
Perkins DJ, Were T, Davenport GC et al (2011) Severe malarial anemia: innate immunity and pathogenesis. Int J Biol Sci 7:1427–1442

Phillips RE, Looareesuwan S, Warrell DA et al (1986) The importance of anaemia in cerebral and uncomplicated falciparum malaria: role of complications, dyserythropoiesis and iron sequestration. Q J Med 58:305–323

Piña-Vázquez C, Reyes-López M, Ortíz-Estrada G et al (2012) Host-parasite interaction: parasite-derived and -induced proteases that degrade human extracellular matrix. J Parasitol Res 2012:748206. doi:10.1155/2012/748206

Pino P, Taoufiq Z, Nitcheu J et al (2005) Blood–brain barrier breakdown during cerebral malaria: suicide or murder? Thromb Haemost 94:336–340

Polimeni M, Prato M (2014) Host matrix metalloproteinases in cerebral malaria: new kids on the block against blood–brain barrier integrity? Fluids Barriers CNS 11:1. doi:10.1186/2045-8118-11-1

Pouvelle B, Buffet PA, Lépolard C (2000) Cytoadhesion of Plasmodium falciparum ring-stage-infected erythrocytes. Nat Med 6:1264–1268

Prato M, Giribaldi G (2011) Matrix metalloproteinase-9 and haemozoin: wedding rings for human host and plasmodium falciparum parasite in complicated malaria. J Trop Med 2011:628435. doi:10.1155/2011/628435

Prato M, Giribaldi G, Polimeni M et al (2005) Phagocytosis of hemozoin enhances matrix metalloproteinase-9 activity and TNF-alpha production in human monocytes: role of matrix metalloproteinases in the pathogenesis of falciparum malaria. J Immunol 175:6436–6442

Prato M, Gallo V, Giribaldi G et al (2008) Phagocytosis of haemozoin (malarial pigment) enhances metalloproteinase-9 activity in human adherent monocytes: role of IL-1beta and 15-HETE. Malar J 7:157. doi:10.1186/1475-2875-7-157

Prato M, Gallo V, Valente E et al (2010a) Malarial pigment enhances Heat Shock Protein-27 in THP-1 cells: new perspectives for in vitro studies on monocyte apoptosis prevention. Asian Pac J Trop Med 3:934–938

Prato M, Gallo V, Arese P (2010b) Higher production of tumor necrosis factor alpha in hemozoin-fed human adherent monocytes is dependent on lipidic component of malarial pigment: new evidences on cytokine regulation in Plasmodium falciparum malaria. Asian Pac J Trop Med 3:1–5

Prato M, Khadjavi A, Mandili G et al (2011) Insecticides as strategic weapons for malaria vector control. In: Perveen F (ed) Insecticides—advances in integrated pest management, 1st edn. Intech, Croatia, pp 91–114

Price RN, Simpson JA, Nosten F (2001) Factors contributing to anemia after uncomplicated falciparum malaria. Am J Trop Med Hyg 65:614–622

Roberts L, Enserink M (2007) Did they really say… eradication? Science 318:1544–1545

Rogier C, Hommel M (2011) Plasmodium life-cycle and natural history of malaria. In: Impact malaria sanofi Aventis' commitment. http://www.impact-malaria.com/iml/cx/en/layout.jsp?cnt=2FAECA4CCA72-4C97-9090-B725867E1579

Rowe JA, Claessens A, Corrigan RA et al (2009) Adhesion of Plasmodium falciparum-infected erythrocytes to human cells: molecular mechanisms and therapeutic implications. Expert Rev Mol Med 11:e16. doi:10.1017/S1462399409001082

Salmon BL, Oksman A, Goldberg DE (2001) Malaria parasite exit from the host erythrocyte: a two-step process requiring extra erythrocytic proteolysis. Proc Natl Acad Sci U S A 98:271–276

Scherf A, Pouvelle B, Buffet PA (2001) Molecular mechanisms of Plasmodium falciparum placental adhesion. Cell Microbiol 3:125–131

Schwarzer E, Turrini F, Ulliers D et al (1992) Impairment of macrophage functions after ingestion of Plasmodium falciparum infected erythrocytes or isolated malarial pigment. J Exp Med 176:1033–1041

Scorza T, Magez S, Brys L (1999) Hemozoin is a key factor in the induction of malaria-associated immunosuppression. Parasite Immunol 21:545–554

Slater AF (1992) Malaria pigment. Exp Parasitol 74:362–365. doi:10.1016/0014-4894(92)90162-4

Slater AF, Swiggard WJ, Orton BR et al (1991) An iron-carboxylate bond links the heme units of malaria pigment. Proc Natl Acad Sci U S A 88:325–329

Szklarczyk A, Stins M, Milward EA et al (2007) Glial activation and matrix metalloproteinase release in cerebral malaria. J Neurovirol 13:2–10

Tan KR, Arguin PM, Steele SF (2014) Travel vaccines & malaria information, by country in CDC health information for international travel (chapter 3). Centers for Disease Control and Prevention, Atlanta, http://wwwnc.cdc.gov/travel/yellowbook/2014/chapter-3-infectious-diseases-related-to-travel/travel-vaccines-and-malaria-information-by-country

Taylor WR, Cañon V, White NJ (2006) Pulmonary manifestations of malaria: recognition and management. Treat Respir Med 5:419–428

Taylor WR, Hanson J, Turner GD (2012) Respiratory manifestations of malaria. Chest 142: 492–505. doi:10.1378/chest.11-2655

Tilley L, Dixon MW, Kirk K (2011) The Plasmodium falciparum-infected red blood cell. Int J Biochem Cell Biol 43:839–842. doi:10.1016/j.biocel.2011.03.012

Urban BC, Todryk S (2006) Malaria pigment paralyzes dendritic cells. J Biol 5:4

Van den Steen PE, Van Aelst I, Starckx S et al (2006) Matrix metalloproteinases, tissue inhibitors of MMPs and TACE in experimental cerebral malaria. Lab Invest 86:873–888

Van den Steen PE, Deroost K, Deckers J et al (2013) Pathogenesis of malaria-associated acute respiratory distress syndrome. Trends Parasitol 29:346–358. doi:10.1016/j.pt.2013.04.006

Weber MW, Böker K, Horstmann RD et al (1991) Renal failure is a common complication in non-immune Europeans with Plasmodium falciparum malaria. Trop Med Parasitol 42:115–118

Were T, Davenport GC, Hittner JB et al (2011) Bacteremia in Kenyan children presenting with malaria. J Clin Microbiol 49:671–676

Wheeler AP, Bernard GR (2007) Acute lung injury and the acute respiratory distress syndrome: a clinical review. Lancet 369:1553–1564

White NJ (1998) Not much progress in treatment of cerebral malaria. Lancet 352:594–595

WHO (2000) Severe falciparum malaria; World Health Organization, communicable diseases cluster. Trans R Soc Trop Med Hyg 94:S1–S90

WHO (2013) World malaria report: 2013. WHO library cataloguing-in-publication data. http://apps.who.int/iris/bitstream/10665/97008/1/9789241564694_eng.pdf

William T, Menon J, Rajahram G et al (2011) Severe Plasmodium knowlesi malaria in a tertiary care hospital, Sabah, Malaysia. Emerg Infect Dis 17:1248–1255

Chapter 2
Malaria Diagnosis, Therapy, Vaccines, and Vector Control

Nicoletta Basilico, Roberta Spaccapelo, and Sarah D'Alessandro

1 Introduction

Depending on the transmission intensities and the age of the patient, human malaria manifests with different clinical outcomes ranging from asymptomatic malaria, mild uncomplicated disease to life-threatening severe disease. In addition, different vectors can be associated with different parasite strains, thereby further influencing the clinical outcomes. In this context, it appears evident that diagnosis, therapy, vaccine, and vector control approaches must be multiple. The present chapter will review the current state of the art of the approaches currently employed to counteract the various manifestations of malaria.

2 Malaria Diagnosis

An effective management and treatment of malaria depends on early and accurate diagnosis, as misdiagnosis can result in significant morbidity and mortality. In particular, it is essential to diagnose the presence of *P. falciparum* early since the infection can be fatal within few days.

N. Basilico (✉)
Dipartimento di Scienze Biomediche, Chirurgiche e Odontoiatriche, Università degli Studi di Milano, Milan, Italy
e-mail: nicoletta.basilico@unimi.it

R. Spaccapelo
Dipartimento di Medicina Sperimentale, Università degli Studi di Perugia, Perugia, Italy

S. D'Alessandro
Dipartimento di Scienze Farmacologiche e Biomolecolari, Università degli Studi di Milano, Milan, Italy

M. Prato (ed.), *Human and Mosquito Lysozymes*,
DOI 10.1007/978-3-319-09432-8_2

The diagnosis of malaria is sought on clinical signs and symptoms and is based on the detection of parasites in the blood (parasitological diagnosis). The first symptoms of malaria are nonspecific (headache, muscle pains, fatigue, fever, chills, nausea, vomiting) and similar to several viral illnesses. In many endemic areas, malaria is therefore frequently overdiagnosed. World Health Organization (WHO) recommends rapid parasitological confirmation of diagnosis (by microscopy or malaria rapid diagnostic test [RDT]) in all patients with suspected malaria prior to the start of treatment (WHO 2013). This is important either to restrict use of antimalarial drugs only to the patients with malaria or to reduce the emergence and spread of drug resistance.

2.1 *Microscopic Diagnosis*

Microscopic slide examination of peripheral blood is regarded as the "gold standard" for malaria diagnosis and remains the most widely used test in most endemic areas. When used appropriately, microscopy is inexpensive, rapid, and relatively sensitive offering significant advantages over other methods. Sensitivity is about 10 parasites/μL and about 500 parasites/μL for thick film and thin film, respectively. Although thick film is more sensitive for detecting the presence of parasite, thin film can be used for accurate identification of species (except for *P. knowlesi* infection) and for counting the parasitemia. Microscopy offers indeed the advantage to identify the presence of mixed infections and can also be used to monitor the efficacy of therapy by successive examination of the parasitemia until the complete clearance (Kilian et al. 2000; Mayxay et al. 2004). However, microscopic slide examination might not be able to detect very low parasitemias; moreover, it is time consuming and needs adequately trained and skilled people while reading the smears to make accurate and reproducible diagnoses.

Thick blood films consist of a thick layer of lysed red blood cells (RBCs) since it is stained as unfixed preparation using diluted Giemsa or Wright's stain (see Table 2.1). The thin blood film is methanol fixed, thus the morphological identification of the parasite species is much easier.

A standard way of estimating the parasite count per μL of blood on thick film is to use a standard value for the leucocytes count (8,000 WBC/μL). Counting the number of parasites present in thick film until 200 white blood cells (WBCs) has been seen and then multiplying the counted parasites by 40 will give the number of parasite count per microliter of blood (Warhurst and Williams 1996). Quantification should also be performed using a thin film by counting the number of parasitized RBCs (not individual parasite) per 100 RBCs in a thin film (at least 1,000 RBCs have to be counted).

Different morphological characteristics of both parasites and infected RBCs should be considered for accurate species identification in a thin film (Table 2.1).

In general, all stages of the parasite can be seen in the case of *P. vivax*, *P. ovale*, and *P. malariae*, but usually only the ring parasites (see Fig. 2.1) and the banana-like

Table 2.1 Diagnosis of *Plasmodium* species in Giemsa stained thin film

	P. falciparum	*P. vivax*	*P. malariae*	*P. ovale*
Ring forms	Always present in peripheral blood. Small-medium size. Multiple infection, marginal position, double chromatin dots common	Large cytoplasm, large chromatin dot. Multiple infections not common	Thick cytoplasm, large chromatin dot. Multiple infections not common	Round to oval. Thick cytoplasm, large chromatin dot. Multiple infections not common
Trophozoites Growing form	Rarely present (only in severe infections). RBCs unaltered in size. Compact cytoplasm, dark pigment	Enlarged RBCs. Large and ameboid parasite. Yellowish-brown pigment	RBCs unaltered in size. Parasite compact, occasionally band form. Dark-brown pigment	RBCs unaltered or slightly enlarged in size, some oval and fimbriated. Dark-brown pigment
Schizonts	Rarely present (only in severe infections). RBCs unaltered in size. 8–24 merozoites and dense black pigment clumped in one mass	Normally present in peripheral blood. RBCs enlarged. Large parasite. 12–24 merozoites. Yellowish-brown pigment	Normally present in peripheral blood. RBCs oval, fimbriated. 6–14 merozoites with large nuclei. Dark-brown, central pigment	Normally present in peripheral blood. RBCs unaltered in size. 6–12 merozoites with large nuclei, sometimes forming a rosette. Central mass of dark brown pigment
Gametocytes	RBCs distorted by parasite. Parasites are with crescent shape. Chromatin in a single mass in macrogametocyte or diffuse in microgametocyte	RBCs enlarged in size. Large, round parasites. Chromatin compact in macrogametocyte or diffuse in microgametocyte	RBCs normal in size. Round parasites. Chromatin compact in macrogametocyte or diffuse in microgametocyte	RBCs slightly enlarged. Round parasites. Chromatin compact in macrogametocyte or diffuse in microgametocyte
Age of infected erythrocytes	Young and old	Young	Old	Young

RBCs red blood cells

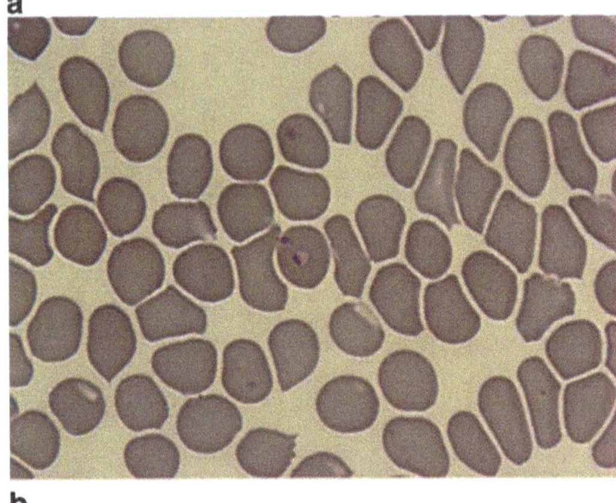

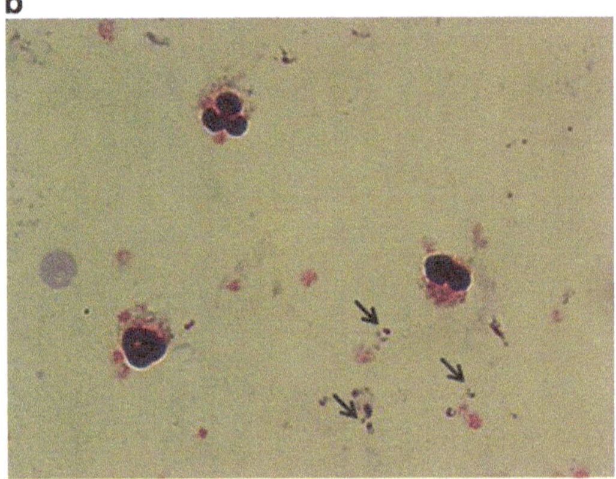

Fig. 2.1 *Plasmodium falciparum*: ring stage trophozoites in a thin (**a**) and a thick (**b**) blood smear. Rings are marked with *arrows* (Courtesy of Dr. Romualdo Grande, Azienda Ospedaliera L. Sacco, Milan, Italy)

gametocytes are distinguished in peripheral blood in the case of *P. falciparum* malaria since mature parasites are sequestered.

Microscopy alone is insufficient to diagnose *P. knowlesi*, forms of which can be confused with *P. malariae* and a PCR test is required to confirm the infection. With high parasitemia *P. malariae*-like parasites should be treated for *P. knowlesi* infection.

The presence of malaria pigment in leukocytes (neutrophils and monocytes) is a prognostic marker of severe malaria (Nguyen et al. 1995).

2.2 *Rapid Diagnostic Test*

Malaria RDTs were developed in the mid-1990s (Dietze et al. 1995; Palmer et al. 1998) and are based on the detection of antigens or enzymatic activities associated with the parasites. RDTs are quick (provide results within 20 min), simple, and easy to perform. A blood sample collected from the patient is placed in a test strip;

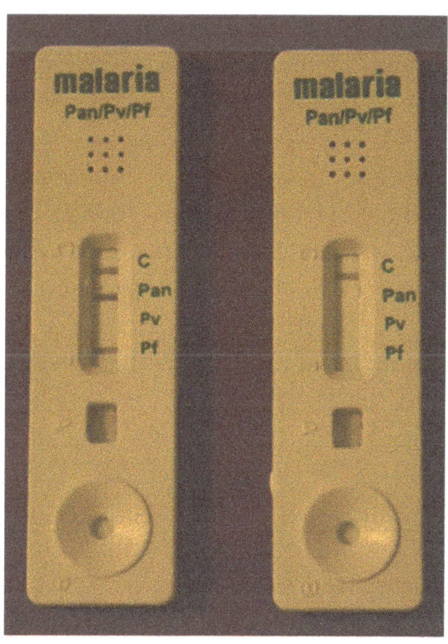

Fig. 2.2 RDTs positive or negative for *P. falciparum*

Table 2.2 Parasite species and target antigen of RDTs

	Target antigen		
	HRP2	pLDH	Aldolase
Species of parasites detected	*P. falciparum* specific	*P. falciparum* specific	Pan specific
		P. vivax specific	
		Pan specific	

Modified from Wilson (2012)

dye-labeled antibodies bind to parasite antigen, the resultant complex migrates on a nitrocellulose strip and is arrested by a capture antibody, forming a visible line. The presence of specific bands in the test card window indicates whether the patient is infected and a control line gives information on the integrity of the antibody–dye conjugate (see Fig. 2.2). RDTs are very useful in endemic areas since they are simple and quick to perform, easily transportable, and they do not require a source of electricity, thus offering a useful alternative to microscopy. Moreover, in non-endemic areas, they can be a complementary test in case of microscopy inexperience.

Some RDTs can detect only one species of malaria parasite, some other can detect more than one. The most common antigens for RDTs are *P. falciparum* histidine-rich protein-2 (*Pf*HRP2), specific for *P. falciparum*, and two enzymes of the parasite glycolytic pathways, namely plasmodial lactate dehydrogenase (pLDH) and aldolase. LDH can be specific for *P. falciparum* or *P. vivax* or it can be a variant pan specific (common to all species: *P. falciparum*, *P. vivax*, *P. malariae*, *P. ovale*, and *P. knowlesi*); aldolase is pan specific (see Table 2.2).

Another advantage in the use of RDTs is that parasite antigens, in particular *Pf*HRP2, can also be measured when mature parasites are sequestered. Sequestration of mature trophozoites and schizonts in the microvasculature of vital organs is a central feature in the pathogenesis of *falciparum* malaria. Peripheral blood parasitemia counted by microscope does not take into account sequestered parasitized RBCs. However, these parasites secrete *Pf*HRP2 into the plasma and thus plasma concentration of this protein can provide a better estimation of the total parasite biomass (Dondorp et al. 2005b). Moreover, it is possible to measure antigens when parasite detection by microscopy is difficult because most of them are sequestered and present at very low parasitemia in the peripheral blood.

The disadvantage in the use of the RDTs is related to the multiplicity of different commercial products with different quality control and variability in the specificity and sensitivity (WHO 2011). The WHO and the Foundation for Innovative New Diagnostics (FIND) offer a testing program to evaluate the performance of RDTs (WHO 2011). Moreover, RDTs suffer of several diagnostic limitations: (1) antigens specific for *P. ovale*, *P. malariae*, or *P. knowlesi* are missing; (2) they cannot be used to determine the parasitemia or to follow the therapy since the malaria antigens (in particular *Pf*HRP2) can persist in the blood after the parasite clearance or are produced also by gametocytes, which are not usually killed by the standard antimalarial therapy; (3) persistence of the *Pf*HRP2 antigen in the blood after parasite clearance can lead to false positives (Iqbal et al. 2002; Mayxay et al. 2001; Mharakurwa and Shiff 1997; Mueller et al. 2007); (4) some isolates from the Amazon region and from Africa and India have been found lacking the *Pf*HRP2 (Gamboa et al. 2010; Houzé et al. 2011; Kumar et al. 2013), which means that RDTs based on this antigen cannot be useful in these regions.

2.3 New Malaria Diagnostic Targets

Some alternative malarial diagnostic targets have been explored and include proteins whose genes are highly conserved (Mouatcho and Goldring 2013).

Heat-shock proteins (HSPs) are produced at high levels by the parasite in response to increases in body temperature during malaria fever (Sharma 1992). Among HSPs, HSP-70—which is most abundant—has been explored as a new diagnostic target. Elevated levels of anti-*Pf*HSPs IgM and IgG have been found in the serum of malaria patients (Minota et al. 1988; Zhang et al. 2001). A *P. vivax* recombinant HSP-70-based ELISA has been shown to detect anti-*Pv*-HSP70 antibodies in the plasma of *P. vivax* and *P. falciparum* malaria with a sensitivity of 88.8 % and 78.8 % for *P. vivax* and *P. falciparum*, respectively (Na et al. 2007).

Another possible diagnostic target antigen could be the *Plasmodium* heme detoxification protein (HDP), a novel identified protein that binds heme and converts it into nontoxic hemozoin (Jani et al. 2008). HDP is present in all the species of *Plasmodium*, is functionally conserved, and seems to be critical for the survival of the parasite (Jani et al. 2008; Vinayak et al. 2009). Monoclonal Abs against *Pf*HDP have been developed and their specificity and affinity were comparable to commercially available histidine-rich protein antibodies, making them promising for further test development (Kattenberg et al. 2012).

Dihydrofolate reductase (DHFR) is an enzyme of the folate pathway that catalyzes the reduction of dihydrofolate to tetrahydrofolate. Unlike bacteria and higher eukaryotes, protozoan DHFR forms a bifunctional enzyme with thymidylate synthase (TS) (Sherman 1979). Moreover, the plasmodial DHFR domain contains two additional regions whose sequences vary slightly in all plasmodia and these differences can therefore be assessed for *Plasmodium* species differentiation (Yuvaniyama et al. 2003). Monoclonal Abs against this protein have also been developed and it has been shown that they can detect *P. falciparum* and *P. vivax* (Kattenberg et al. 2012).

Glutamate-rich protein (GLURP), a highly antigenic exoantigen of *P. falciparum* whose gene is conserved in different isolates, is another possible target only suitable for *P. falciparum* diagnosis, since the gene is not present in the other species. This protein has also been considered as a malaria vaccine candidate (Jepsen et al. 2013).

An alternative approach could be to detect host markers rather than parasite proteins or metabolites.

High mobility group box 1 (HMGB1) protein is an important mediator of inflammation implicated in sepsis pathophysiology. Elevated plasma HMGB1 levels were significantly associated with severe malaria in pediatric patients with *P. falciparum* infection suggesting that elevated HMGB1 could be a prognostic marker of disease severity (Higgins et al. 2013).

Several biomarkers have been proposed for detecting malaria-associated inflammation and infection during pregnancy. *Plasmodium falciparum*-infected erythrocytes sequester in the placenta causing placental malaria, a condition associated with reduced birth weight baby and reduced chance of surviving the first year of life. Since infected erythrocytes are not observed in peripheral blood smears, women with placental malaria are misdiagnosed and thus not treated. Detection of inflammatory mediators in the peripheral blood may provide an approach to diagnose placental malaria. A study from Cameroon reported an association between plasma-soluble TNF receptor-2 levels and low birth weight babies in women infected with *P. falciparum*, suggesting that biomarkers in peripheral blood might discriminate women with poor pregnancy outcomes as a function of infection (Thévenon et al. 2010).

Another study in Tanzania showed that in women infected with *P. falciparum* the levels of IL-10 and IP-10 increased significantly while those of RANTES decreased significantly concluding that this condition might be predictive of *P. falciparum* infections (Boström et al. 2012).

2.4 Molecular Diagnosis

PCR-based methods to detect malaria infection were described in the early 1990s and have proven to be the most sensitive test able to identify low levels of infection, parasite species, or mixed infections (Barker et al. 1992; Snounou et al. 1993).

Some modified PCR methods, e.g., nested PCR, real-time PCR, and reverse transcription PCR, have been developed to augment the sensitivity and specificity.

Recently, the PCR method has become widely accepted for identifying *P. knowlesi* infections since the current microscopic examination may fail to distinguish *P. knowlesi* from *P malariae*, due to their similar morphology (Oddoux et al. 2011).

However, the utility of PCR is limited by high cost, inadequate laboratory facilities, the need for trained operators, and the time to obtain results that is too long to be appropriate in establishing the diagnosis of malaria. Even in non-endemic countries, PCR is not a suitable method for routine use, but it is more useful for confirming the parasite species after the diagnosis has been established by microscopy or RDT. Moreover, PCR is useful as a research tool in clinical trials, epidemiological studies, and for detection of molecular markers of drug resistance to antimalarial drugs.

Loop-mediated isothermal amplification (LAMP) is a simple, inexpensive, specific nucleic acid amplification method, the application of which for malaria diagnosis was reported in 2006 (Poon et al. 2006). A species-specific LAMP diagnostic method for the detection of the four human species was described in 2007 (Han et al. 2007) and more recently the method has been applied to the detection of *P. knowlesi* infection (Iseki et al. 2010; Lau et al. 2011).

LAMP method can be used under field conditions since it requires simple laboratory devices both for amplification and for detection of the target gene. The whole amplification reaction takes place continuously under isothermal conditions (65 °C), thus a simple water bath can be enough. LAMP results can also be carried out with naked eye observation in the form of either visual turbidity or visual fluorescence.

Alternatively, the amplified products can also be visualized using a fluorescent intercalating dye; when SYBR green is employed, the original orange color of the dye change into green, in case of positive amplification. The use of the UV light can increase the fluorescence intensity.

3 Malaria Treatment

Malaria (mainly due to *P. falciparum*) can lead to fatal outcomes in only few days, thus treatment should be started as soon as possible. The main targets of current antimalarial chemotherapy are the asexual blood stages of the parasite, responsible for the malaria symptoms. However, with the new goal of malaria eradication (Khadjavi and Prato 2010; Partnership 2008), the strategy has been changed and the ideal drug should prevent both disease transmission and the relapse of dormant liver stages of *P. vivax and P. ovale*. Drugs able to kill both gametocytes, responsible for malaria transmission from human host to mosquito vector, and hypnozoites, responsible for *P. vivax* and *P. ovale* relapse, are indeed urgently needed. Primaquine represents the only antimalarial currently in therapeutic use able to affect both the mature stage V gametocytes and the hypnozoites. However, primaquine may cause hemolysis in patients with glucose-6-phosphate dehydrogenase (G6PD)-deficiency and is contraindicated in pregnancy and in young children.

Another great challenge to malaria treatment is parasite resistance to almost every class of antimalarial compounds. The use of two or more drugs with different mechanism of action in combination is now recommended for the treatment of *P. falciparum* malaria to delay development of resistance.

3.1 Treatment of Uncomplicated Malaria

WHO recommends artemisinin-based combination therapies (ACTs) for the treatment of uncomplicated malaria caused by *P. falciparum* parasite. ACT is the combination of a fast acting artemisinin derivative (dihydroartemisinin, artesunate, artemether) and a partner drug (amodiaquine -AQ-, mefloquine, piperaquine, lumefantrine) with a different mechanism of action and longer half-life than artemisinins. Almost all countries in which *falciparum* malaria is endemic have adopted ACT as the first-line treatment policy. Unfortunately, fake antimalarials are widespread in African and Asian countries compromising effectiveness and increasing the risk for emergence of resistance (White et al. 2014). Second-line treatment for uncomplicated malaria (to be used when ACT is not available) is a combination of artesunate or quinine and doxycycline, tetracycline, or clindamycin. Quinine plus clindamycin is recommended for treatment of malaria in the first trimester of pregnancy since the safety of artemisinin derivatives during this period is not yet established (WHO 2010).

Chloroquine (CQ) in combination with primaquine for radical cure is still the drugs of choice to treat *P. vivax* malaria in endemic areas where CQ is already effective. However, ACTs should be used in areas where CQ-resistant *P. vivax* parasite has been identified.

3.2 Treatment of Severe Malaria

Severe malaria is a life-threatening disease and should be promptly treated with parenteral antimalarial therapy. Severe malaria is most commonly caused by infection with *P. falciparum*, although *P. vivax* and *P. knowlesi* can also be responsible (Antinori et al. 2013; Price et al. 2009). Over the years, quinine has been the mainstay in the treatment of severe malaria. More recently, injectable artesunate (intramuscular or intravenous) has become the recommended treatment of choice for severe malaria worldwide (WHO 2010), including severe infection caused by *P. vivax* and *P. knowlesi* (Barber et al. 2013). Two large multicenter clinical trials (the first on adult patients in Asia and the second on children in Africa) compared parenteral treatment of artesunate versus quinine in patients with severe malaria revealing that artesunate treatment reduces both adult and child mortality rates (Dondorp et al. 2005a, 2010).

3.3 Antimalarial Drugs

4-Aminoquinolines

CQ, a 4-amino 7-chloro quinoline, is the lead compound of the 4-aminoquinolines family (Fig. 2.3). It was first synthesized in 1934 and for several decades it was the most widely used antimalarial drug. The success of this drug is due to high efficacy, low toxicity, and low-cost synthesis. However, uncontrolled, massive, and often improper treatment with CQ potentiated the emergence and spread of parasite resistance. Nowadays, the use of this drug to treat *falciparum* malaria is limited to few areas even though it still maintains considerable efficacy for the treatment of *P. vivax*, *P. ovale*, and *P. malariae* infections (WHO 2010).

The mechanism of action of CQ is not completely understood. CQ and related 4-aminoquinolines are weak bases which diffuse passively through membranes along pH gradients as unprotonated molecules, and accumulate intracellularly where they are trapped in acidic compartments, become protonated, and raise the local pH (O'Neill et al. 2006). In *Plasmodia*, they accumulate in the parasite food vacuole, where host hemoglobin is degraded to peptides and free heme, which is in turn detoxified by forming inert hemozoin. The antimalarial activity of 4-aminoquinolines is ascribed to their ability to form drug-heme adducts and

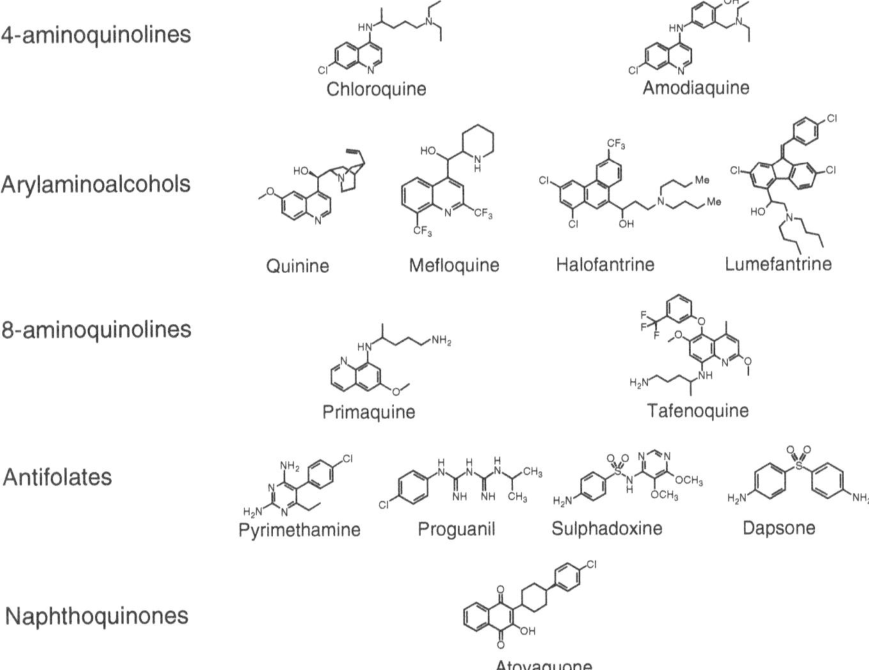

Fig. 2.3 Chemical structures of some current antimalarial drugs

accumulation of free heme, which is toxic for the parasite (Egan 2008; Ridley et al. 1997). Additional effects on parasites have also been reported, such as the inhibition of hemoglobin endocytosis and digestion or disruption of normal vesicle trafficking (Fitch et al. 2003; Hoppe et al. 2004; Roberts et al. 2008).

CQ resistance is related to point mutations in the gene encoding for the *P. falciparum* chloroquine resistance transporter (*Pf*CRT) protein, resulting in reduced CQ accumulation in the food vacuole (Bray et al. 2005; Wellems and Plowe 2001).

AQ is a phenyl-substituted analog of CQ (see Fig. 2.1) with a mechanism of action similar to CQ, but still divisive (O'Neill et al. 1998). Although cross-resistance of CQ and AQ has been documented, AQ is effective against several CQ-resistant strains. However, its clinical use has been severely restricted because of hepatotoxicity and agranulocytosis.

Piperaquine is a bisquinoline antimalarial, first synthesized in the 1960s, with an excellent activity on CQ-resistant parasites (Raynes 1999; Vennerstrom et al. 1992). The exact mechanism of action of piperaquine is unknown hitherto. Nonetheless, it seems to act in a similar way to other quinolines, interfering with the heme detoxification process (Davis et al. 2005). Piperaquine has been used extensively in China as mass prophylaxis and treatment until the emergence of resistance, which diminished its use by the late 1980s. Subsequently, due to its good tolerability, pharmacokinetic profile, and low cost, piperaquine was rediscovered as a promising partner drug for ACT. A fixed oral dose combination of piperaquine and dihydroartemisinin (Eurartesim®) is now available for the treatment of uncomplicated *falciparum* malaria in adults and children.

Aryl Amino Alcohol

Quinine, a member of the cinchona alkaloid family (Fig. 2.3), is one of the oldest antimalarial agents. Cinchona bark extracts were identified as early as 1600 to be effective in treating the tertian fever of malaria. In 1820, quinine was isolated and named by Pierre Joseph Pelletier and Joseph Caventou and remained the mainstay of malaria treatment until the 1920s, when more effective synthetic antimalarials, such as CQ, became available. Quinine was successfully used in mass campaigns that, together with other control measures, led to the eradication of malaria in Italy at the start of the twentieth century (Brown 1997).

Quinine has rapid schizonticidal activity against intraerythrocytic malaria parasites. Its mechanism of action has not been fully elucidated. As the quinoline antimalarials, the most widely accepted hypothesis is that the drug can inhibit hemozoin crystallization interfering with the heme detoxification process.

Although the drug has been used since the seventeenth century, resistance was first reported in 1910 after hundreds of years of use, differently from the resistance to CQ that emerged within only 12 years of its introduction (Peters 1982). Moreover, resistance to quinine is usually in low grade. It has been well documented in Asia (Mayxay et al. 2007) and South America (Legrand et al. 2008), whereas conflicting results have been reported in Africa (Mutanda 1999; Pradines et al. 1998;

Tinto et al. 2006; Touré et al. 2008). Documentation of quinine resistance has been decreasing in this century due to the increasing use of artemisinin.

Quinine has been the mainstay for treating severe malaria since the results of the SEAQUAMAT and AQUAMAT trials have shown the superiority of artesunate with respect to quinine (Dondorp et al. 2005a, 2010). Quinine is now a second line of choice when artesunate is not available.

Mefloquine, a 4-methanolquinoline (Fig. 2.3), was developed at the Walter Reed Army Institute of Research in the 1970s in response to concerns regarding the increase of CQ-resistant parasite. Mefloquine is a blood schizonticide, active against the erythrocytic stages of all malaria species that infect humans, including *P. knowlesi* (Bronner et al. 2009). Mefloquine, atovaquone/proguanil, and doxycycline are all medications licensed and recommended for malaria chemoprophylaxis to areas of chloroquine-resistant *P. falciparum.* Mefloquine is indeed effective in the prevention of malaria, except in clearly defined Thai border regions of multidrug resistance. The mechanism of action of mefloquine is still unknown, probably being different from 4-aminoquinolines. Activity on the parasite seems to be related to the ability of mefloquine to interfere with the transport of hemoglobin from the erythrocyte to the food vacuole (Olliaro 2001).

Halofantrine and lumefantrine are other two compounds of this class, on the market alone or in combination with an artemisinin derivative, respectively. Lumefantrine is available only in an oral preparation co-formulated with artemether.

8-Aminoquinolines

8-Aminoquinolines (Fig. 2.3) are an important class of antimalarials because they belong to the only class proven to be effective against the hypnozoites of *P. vivax* and *P. ovale*. Primaquine is certainly used to achieve radical cure (complete elimination) of relapsing malaria due to *P. vivax* or *P. ovale*, in combination with a blood schizontocide for the erythrocytic parasites. In addition to the activity against hypnozoites, 8-aminoquinolines can kill gametocytes and consequently block the malaria transmission. The addition of a single dose of primaquine to ACT treatment is, therefore, recommended by WHO to reduce gametocyte burden and thus transmission. G6PD deficiency should be evaluated before the administration of primaquine since the most significant adverse effect is hemolytic anemia in patients with G6PD deficiency. Primaquine is not recommended for pregant women and for children under 4 years (WHO 2010). The mechanism of action of primaquine is unknown but the mitochondria may be the biological target. It is thought to interfere with the cellular respiration of the parasite by means of generating oxygen-free radicals and deregulating the electron transport (Krungkrai et al. 1999).

Tafenoquine is a primaquine analog developed at the Walter Reed Army Institute of Research in 1979 with aim to find less toxic and longer acting 8-aminoquinolines. Tafenoquine possesses higher activity than primaquine both in in vitro assays on *P. falciparum* and in in vivo animal models with *P. berghei* (Coleman et al. 1992;

Vennerstrom et al. 1999). Moreover tafenoquine retains gametocytocidal and sporontocidal activities representing a hopeful candidate agent for transmission-blocking public health applications.

Antifolates

Antifolates (see Fig. 2.3) are drugs able to block the synthesis or the conversion of folate derivatives, thus indirectly they have an effect on the synthesis of nucleic acids in the malaria parasite.

The antifolates currently in use can target two important enzymes of the folate pathway, namely the DHFR and dihydropteroate synthase (DHPS). Pyrimethamine and proguanil target DHFR, whereas sulfadoxine and dapsone act on DHPS. These drugs have not been used for long time alone since it was demonstrated that DHPS inhibitors could synergize with inhibitors of DHFR (Bushby 1969) leading to their use in combination. Proguanil, the first antifolate to be discovered, is converted in the liver into the active metabolite, cycloguanil. It is not actually used alone as resistance to proguanil develops very quickly. The combination of proguanil and atovaquone, an inhibitor of the electron transport, is known as Malarone® and it is used for both treatment and prophylaxis.

Pyrimethamine is used in synergistic combination with sulfadoxine (Fansidar®) or sulfalene (Metakelfin®) for treatment of uncomplicated malaria and with dapsone for prophylaxis.

Pyrimethamine/sulfadoxine has been the drug of choice for intermittent preventive treatment (IPT) in pregnant women living in malaria-endemic areas. It has been shown that the administration of one dose of pyrimethamine/sulfadoxine in the second and third trimester of pregnancy is associated with a reduction of placental parasitemia, maternal anemia, and low birth weight (Grobusch et al. 2007).

Resistance, caused by point mutations in DHFR and DHPS, arises rather rapidly in response to drug pressure and is now common worldwide.

Atovaquone is a hydroxynaphthoquinone antiparasitic drug active against all *Plasmodium* species. Atovaquone selectively inhibits the parasite mitochondrial electron transport chain at the cytochrome bc1 complex. Selectivity is due to structural differences between the cytochrome b encoded by the parasite mitochondrial DNA and that encoded by the host mitochondrial DNA (Vaidya et al. 1993). As stated previously, it is currently used in combination with proguanil for both treatment and prophylaxis. Resistance is related to mutations of cytochrome b gene (Srivastava et al. 1999).

Artemisinin and Derivatives

Artemisinin (quinghaosu) is an endoperoxide sesquiterpene lactone (see Fig. 2.4) extracted from the Chinese herb *qinghao* (*Artemisia annua*), a herbal remedy used in China for the treatment of fever for about 2,000 years. In 1967, the Chinese

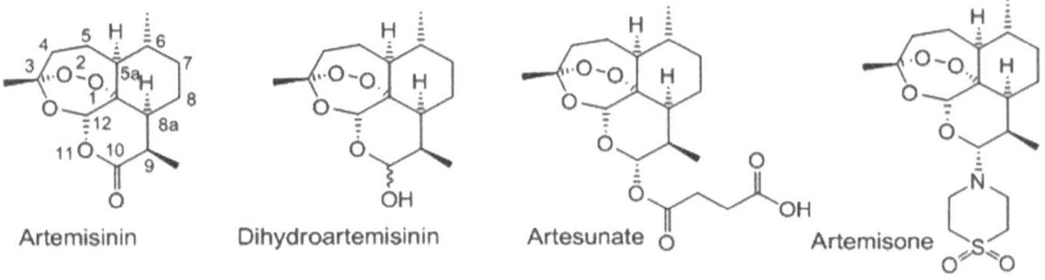

Fig. 2.4 Chemical structures of artemisinin and some first generation (dihydroartemisinin and artesunate) and second generation (artemisone) derivatives

government launched a coordinated program to discover antimalarial compounds in indigenous plants used in traditional Chinese medicine which led to the discovery of quinghaosu (now called artemisinin), an extract of quinghao, with potent antimalarial activity in 1971. Artemisinin is a potent and fast acting blood schizontocide killing all parasite stages including young *P. falciparum* gametocytes (stages I–IV). However, artemisinin has some pharmacokinetic limitations such as low solubility, poor bioavailability, and short half-life in vivo. To overcome some of these problems, semisynthetic derivatives have been developed. First generation derivatives include dihdyroartemisinin, artesunate, arteether, and artemether. All of these compounds share the same basic chemical structure of artemisinin but possess different chemical groups at the C10 position (see Fig. 2.4). Moreover, all active compounds possess a distinctive 1,2,4-trioxane pharmacophore, which is essential for the antimalarial activity since the corresponding acyclic compounds lacking the endoperoxide are biologically inactive (Posner et al. 1992). Second generation artemisinin includes artemisone which has shown improved pharmacokinetic properties such as longer half-life and lower toxicity (Haynes 2006).

Artemisinin derivatives, alone or in combination, are the treatment of choice for severe and uncomplicated malaria, respectively. In particular, artesunate is worldwide recommended for severe malaria, including severe *vivax* malaria and *knowlesi* malaria (Barber et al. 2013).

ACT consists of an artemisinin derivative combined with a long-acting antimalarial drug. Different ACT options are now available and include artemether/lumefantrine, artesunate/AQ, artesunate/mefloquine, artesunate/sulfadoxine-pyrimethamine, and dihydroartemisinin/piperaquine. To promote patient adherence to treatment and to avoid the use of artemisinins as monotherapies, fixed-dose combination formulations into a single tablet are strongly recommended (WHO 2010) and are now available for all recommended ACTs, except for artesunate plus SP (WHO 2010).

Artemisinins should be avoided in the first trimester of pregnancy until more information is available. Morphological abnormalities in early pregnancy have been indeed demonstrated in animal studies (Longo et al. 2006). The precise mechanism of action of artemisinin is unclear and still controversial (O'Neill et al. 2010). It has been suggested that the endoperoxide bond undergoes reductive activation by iron(II) or iron(II)-heme. This redox reaction produces carbon-centered radicals that alkylate target molecules leading to parasite's death (Meshnick et al. 1991; Olliaro et al. 2001). Alternative views suggest that artemisinin inhibits a *P. falciparum* sarcoendoplasmic reticulum Ca^{2+}-ATPase (SERCA) homolog (Eckstein-Ludwig et al. 2003). Another proposed mechanism is that artemisinins act as oxidant drugs through oxidation of FADH2 and parasite flavoenzymes (Haynes et al. 2010, 2012).

Parasite resistance to artemisinin has so far been described in Southeast Asian countries (Cambodia, Myanmar, Thailand, and Viet Nam). Several factors, such as low patient adherence to antimalarial regimens, counterfeit drugs, and monotherapies, contributed to the emergence of artemisinin resistance. If resistances were to spread, the public health consequences could be disastrous considering that no alternative antimalarials, with comparable efficacy to artemisinins, are presently available.

4 Malaria Vaccines

The emergence and spread of artemisinin and insecticide resistance have shown the limitations of the current recommended malaria control measures, thus a safe and protective vaccine is needed to achieve the goals of WHO's malaria eradication agenda. If malaria eradication has to be achieved, not only *P. falciparum* but also *P. vivax* vaccine has to be developed.

The rationale for a malaria vaccine development is based on the observation that people living in endemic areas develop clinical protective immunity that limits the severity of the disease and prevent mortality (Greenwood 2005).

Despite decades of intense research in this field, no licensed malaria vaccines are available until now. The most advanced one in development is the RTS,S subunit vaccine which targets the circumsporozoite protein of *P. falciparum*. The listless development in this field and the complexity in achieving a vaccine are due to the strong ability of malaria parasite to evade host's immune system. During the multistage life cycle, malaria parasites express different antigens, each stimulating a specific immune response. Moreover, parasites exhibit wide genetic diversity, particularly in the surface proteins which could be employed as vaccine antigens; some blood-stage antigens, such as *P. falciparum* erythrocyte membrane protein-1 (*Pf*EMP1), show temporal switching of variant expression. There is also a lack of understanding of the specific immune responses able to confer protection against the parasite in humans.

The main required characteristics of an ideal malaria vaccine should include high efficacy in preventing clinical disease and transmission, good safety profile for young infants and pregnant women, and protection against the five species of malaria parasite or at least against *P. falciparum* and *P. vivax*.

Vaccination against malaria can target different life cycle stages. Vaccines can be indeed divided into three types:

- Preerythrocytic vaccines
- Blood-stage vaccines
- Transmission-blocking vaccines

4.1 Preerythrocytic Vaccines

Target of these vaccines are sporozoites and/or hepatic stages of the parasite. These vaccines should be able to elicit an immune response mediated by an antibody that prevents invasion or by T cells that attack the infected liver cell.

Sterile immunization with whole irradiated sporozoites, inoculated by mosquito bite, was demonstrated both in mice (Nussenzweig et al. 1967) and in humans (Clyde et al. 1973) more than 40 years ago. However, in order to obtain a whole parasite vaccine, sporozoites have to be cryopreserved and then administered via standard methods, such as intramuscular and intradermal injections. Little progress in improving these models has been made until recently. In 2011, Epstein et al. demonstrated that intradermal inoculation of cryopreserved irradiated *P. falciparum* sporozoites to 80 volunteers was safe but neither immunogenic nor protective (Epstein et al. 2011). In contrast five intravenous inoculations of attenuated, cryopreserved *P. falciparum* sporozoites protected all the six participants of the study (Seder et al. 2013). A genetically attenuated parasite has been proposed as an alternative to irradiation: by dual gene deletions, an attenuated *P. falciparum* unable to complete liver stage development has been obtained (VanBuskirk et al. 2009).

The circumsporozoite antigen was identified as the major component of the sporozoite surface and the gene was cloned and sequenced. This antigen reacts with antibodies that inhibit the invasion of hepatocytes by sporozoites and induce T cell responses against sporozoite-infected liver cells (Nardin and Nussenzweig 1993; Rieckmann et al. 1974). These findings led to circumsporozoite antigen development as a candidate vaccine and the RTS,S is now the leading candidate vaccine for malaria. RTS,S is a recombinant subunit vaccine composed of the central repeat region of *P. falciparum* circumsporozoite protein fused to hepatitis B surface antigen (HBsAg). The fused protein is coexpressed in yeast cells with free HBsAg (Gordon et al. 1995). RTS,S antigen construct on its own has limited immunogenicity and it has since been formulated with potent adjuvant system, AS02 and AS01. The results from a phase III trial in children 5–17 months of age showed reduced episodes of both clinical and severe malaria by approximately 50 % (Agnandji et al. 2011). Another large multicenter phase III trial of RTS,S in infants aged 6–12 weeks showed good safety, but moderate efficacy with 30 % protection against clinical malaria and 26 % against severe malaria for 12 months follow-up after the last vaccination (Agnandji et al. 2012). These results show that the vaccine does not meet the ambitious 2015 goal of the Malaria Vaccine Technology Roadmap that a malaria vaccine must provide 50 % protection against severe disease and death for at least 1 year.

4.2 Blood Stage Vaccines

The objectives of these vaccines are preventing the erythrocyte invasion and blocking the adherence of parasitized RBCs to several tissues. They do not prevent infection but attenuate the clinical symptoms of malaria. Among the large number of asexual blood stage vaccine candidates, most target merozoite antigens, such as the merozoite surface protein 1 (MSP1), 2 (MSP2), 3 (MSP3), the apical membrane protein (AMA1), the GLURP, and the erythrocyte-binding antigen 175 (EBA175). None have resulted in clear clinical protection, and generally phase II trials showed limited efficacy (Fowkes et al. 2010; Goodman and Draper 2010; Richards and Beeson 2009).

Antigens expressed on the surface of infected RBCs are highly polymorphic and thus they are not good candidates for vaccine development. An exception is a variant of the erythrocyte membrane protein 1 (*Pf*EMP1) known as VAR2CSA, studied in the search of a vaccine for prevention of pregnancy-associated malaria. This is a variation of *falciparum* malaria in prima gravidas, associated with maternal anemia and placental malaria infection, reducing the birth weight and increasing the risk of neonatal mortality. Pregnancy-associated malaria is characterized by accumulation of infected erythrocytes in the placenta, and this process is mediated by parasite-encoded VAR2CSA binding to chondroitin sulfate A (CSA). The feasibility of a vaccine able to prevent complication during pregnancy is supported by the fact that women acquire immunity after one pregnancy through the production of antibodies against VAR2CSA that inhibits the adhesion of infected erythrocytes to placental CSA (Rogerson et al. 2007). Multigravidae who have acquired antibodies against VAR2CSA are indeed protected from pregnancy-associated malaria. Vaccine candidates directed against VAR2CSA are currently under development (Hviid 2010).

4.3 Transmission-Blocking Vaccine

Transmission-blocking vaccines target the sexual stages of the parasite preventing malaria transmission and thus they will not protect the vaccinated individual but the community by reducing the incidence of infection.

The feasibility of a transmission-blocking vaccine is supported by the observation that sera from immune individuals can block the fertilization of gametes and prevent further parasite development (Greenwood 2005; Greenwood et al. 2008). Oocyst formation and sporogonic cycle in the mosquito midgut can be blocked by specific host antibodies, complement proteins, and cytokines (Sinden 2010; Sutherland 2009). Prevention of sporogonic development can block the transmission of the parasite to the next human and subsequent spread of parasites in endemic populations. Targets of transmission-blocking immunity are surface proteins expressed on gametocytes, gametes, zygotes, and ookinetes. *Pfs*25, *Pfs*28, *Pfs*48/45, and *Pfs*230 of *P. falciparum* have been shown to induce antibodies with transmission-blocking activity when ingested by the mosquito vector during the blood meal (Pradel 2007).

Transmission-blocking vaccine candidates identified in *Plasmodium vivax* are the ookinete surface proteins *Pv*s25 and *Pv*s28 and the gamete surface protein *Pv*s230 (Sattabongkot et al. 2003; Tachibana et al. 2012). Among these leading candidates, *Pf*s 25 and *Pv*s25 have completed Phase I clinical trials with limited results (Wu et al. 2008). Transmission-blocking targets have been also identified in the Anopheles mosquito and include mosquito midgut ligands. The biggest advantage of targeting mosquito ligands is the possibility of obtaining a vaccine able to interrupt the transmission of more than one plasmodial species (Dinglasan and Jacobs-Lorena 2008).

5 Malaria Vector Control

Unlike other infections, and in particular from the other two major global public health threats HIV and TBC, malaria absolutely requires vectors to be transmitted. Vector control is indeed an essential and effective measure of malaria prevention and was responsible for malaria elimination from some countries in the past. Wide-scale use of insecticide-treated nets (ITNs) and indoor residual spraying (IRS) have contributed to the fall in malaria morbidity and mortality in the last decades (WHO 2013). ITNs should be used in all endemic areas; they possess several properties protecting the users (not only against mosquitoes but also against other insects carrying diseases) as well as the community by killing infective insects. IRS is effective for indoor night biters mosquitoes and not for outdoor daytime biters, thus its efficacy is mostly dependent on the insect's behavior.

Pyrethroids are the only insecticides currently recommended by the WHO for use on bed nets (WHO 2013). It is therefore not surprising that during the last decades, pyrethroids-resistant mosquitoes have emerged in some areas of Africa (Ranson et al. 2011).

Three other classes of insecticides, organochlorines (OCs), organophosphates (OPs), carbamates (Cs), have been used in IRS throughout Africa, and resistance to all three had been reported in mosquitoes (WHO 2012).

5.1 *Organochlorines*

OCs belong to the class of chlorinated hydrocarbons, which are compounds that contain chlorine. The most famous compound of this class is Dichlorodiphenyltrichloroethane (DDT) (see Fig. 2.5), a pesticide with a long history of widespread use around the globe. OCs have the advantage to be cheap and very effective; however, they can have serious consequences on environment and on human health. OCs are indeed very stable compounds and can persist in the environment for long periods of time. Since they are lipophilic they can concentrate in body fat tissue and accumulate in animals and humans through the food chain (Kutz et al. 1991).

Fig. 2.5 Chemical structure of DDT

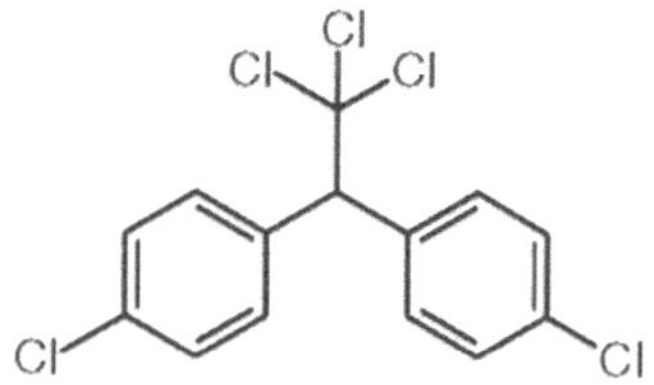

DDT shares the same mechanism of action of pyrethroids, inhibiting the sodium channels. Insects with mutations in sodium channel gene can become resistant to DDT and to other related insecticides. DDT resistance is also conferred by overproduction of detoxification enzyme such as cytochrome P450 through enhanced biodegradation of the insecticide (Müller et al. 2008).

5.2 Organophosphates and Carbamates

OPs and Cs share the same mechanism of action, the inhibition of acetylcholinesterase, an enzyme responsible for neurotransmitter acetylcholine degradation at the cholinergic nerve synapse. Structural differences between the various OPs and Cs affect the efficiency and degree of acetylcholinesterase inhibition. With the spread of pyrethroid resistance, OPs and Cs may be vital alternative insecticides for IRS. However, resistance to OPs and Cs based on reduced sensitivity of acetylcholinesterase has been detected among *An. gambiae* from Cote d'Ivoire (N'Guessan et al. 2003).

5.3 Pyrethroids

Synthetic PYs represent one of the newer classes of insecticides and their use has increased significantly over the last 20 years. Although their chemical structure is different from that of OCs, their toxic effects are similar, acting on the insect's neuronal voltage-gated sodium ion channels in the axonal membranes of insect nerves. Resistance mechanisms are also shared by pyrethroids and DDT and cross-resistance between these compounds restricts the choice of alternative insecticides that can be used to manage resistance.

5.4 Transgenic Mosquitoes

Another approach for controlling transmission could be the use of genetically modified mosquitoes unable to transmit parasites. Several molecules able to block parasite transmission in mosquitoes have been investigated and competitor peptides able to bind salivary gland receptors have been tested. *Anopheles stephensi*, genetically modified for the expression of the SM1 peptide able to bind an ookinete receptor on midgut epithelium, inhibited the ability of mosquito to transmit parasites (Ghosh et al. 2001). Other peptides displaying parasiticidal effects in mosquitoes such as cecropins, defensins, and scorpine have been also described (Gwadz et al. 1989; Kokoza et al. 2010). For a better explanation of these peptides see Chap. 7.

An interesting way to interfere with malaria transmission is the use of paratransgenesis for delivering anti-*Plasmodium* effector molecules by using genetically modified symbiotic microorganism of the insect (Hurwitz et al. 2011).

Lastly, the use of transgenic procedure can ameliorate the sterile insect technique (SIT) for anopheles mosquitoes. SIT approach relies on the release of radiation-sterilized males in the field to compete with wild ones. Currently, sterility in insects is induced by radiation; however, this approach can induce deleterious somatic effects leading to poor mating competitiveness of the sterile male compared to the wild insect (Benedict and Robinson 2003). The use of transgenic approach to induce sterility in male could increase the efficiency of SIT-based strategy (Catteruccia et al. 2009).

However, once a transgenic mosquito is obtained, the next step is its introduction in the field in order to substitute the wild vector population. In order to reach this focal point, several aspects, such as the potentially harmful ecological effects and the acceptance of genetically modified organism usage, have to be carefully investigated.

6 Conclusion

Despite many efforts, the burden of malaria is still great and to reach the goal of malaria elimination and eradication it is necessary to improve all the strategies to combat malaria. These include new diagnostic methods, drugs able to block the transmission of the disease, new classes of insecticides to control mosquitoes and an effective vaccine to prevent infection and/or transmission.

Progresses in each of these approaches could make a huge contribution to malaria eradication.

References

Agnandji ST, Lell B, Soulanoudjingar SS et al (2011) First results of phase 3 trial of RTS, S/AS01 malaria vaccine in African children. N Engl J Med 365:1863–1875

Agnandji ST, Lell B, Fernandes JF et al (2012) A phase 3 trial of RTS, S/AS01 malaria vaccine in African infants. N Engl J Med 367:2284–2295

Antinori S, Galimberti L, Milazzo L et al (2013) Plasmodium knowlesi: the emerging zoonotic malaria parasite. Acta Trop 125:191–201

Barber BE, William T, Grigg MJ et al (2013) A prospective comparative study of knowlesi, falciparum, and vivax malaria in Sabah, Malaysia: high proportion with severe disease from Plasmodium knowlesi and Plasmodium vivax but no mortality with early referral and artesunate therapy. Clin Infect Dis 56:383–397

Barker RH, Banchongaksorn T, Courval JM et al (1992) A simple method to detect Plasmodium falciparum directly from blood samples using the polymerase chain reaction. Am J Trop Med Hyg 46:416–426

Benedict MQ, Robinson AS (2003) The first releases of transgenic mosquitoes: an argument for the sterile insect technique. Trends Parasitol 19:349–355

Boström S, Ibitokou S, Oesterholt M et al (2012) Biomarkers of Plasmodium falciparum infection during pregnancy in women living in northeastern Tanzania. PLoS One 7:e48763

Bray PG, Martin RE, Tilley L et al (2005) Defining the role of PfCRT in Plasmodium falciparum chloroquine resistance. Mol Microbiol 56:323–333

Bronner U, Divis PC, Färnert A et al (2009) Swedish traveller with Plasmodium knowlesi malaria after visiting Malaysian Borneo. Malar J 8:15

Brown PJ (1997) Malaria, miseria, and underpopulation in Sardinia: the "malaria blocks development" cultural model. Med Anthropol 17:239–254

Bushby SR (1969) Combined antibacterial action in vitro of trimethoprim and sulphonamides. The in vitro nature of synergy. Postgrad Med J 45(suppl):10–18

Catteruccia F, Crisanti A, Wimmer EA (2009) Transgenic technologies to induce sterility. Malar J 8(suppl 2):S7

Clyde DF, Most H, McCarthy VC et al (1973) Immunization of man against sporozite-induced falciparum malaria. Am J Med Sci 266:169–177

Coleman RE, Clavin AM, Milhous WK (1992) Gametocytocidal and sporontocidal activity of antimalarials against Plasmodium berghei ANKA in ICR mice and anopheles stephensi mosquitoes. Am J Trop Med Hyg 46:169–182

Davis TM, Hung TY, Sim IK et al (2005) Piperaquine: a resurgent antimalarial drug. Drugs 65:75–87

Dietze R, Perkins M, Boulos M et al (1995) The diagnosis of Plasmodium falciparum infection using a new antigen detection system. Am J Trop Med Hyg 52:45–49

Dinglasan RR, Jacobs-Lorena M (2008) Flipping the paradigm on malaria transmission-blocking vaccines. Trends Parasitol 24:364–370

Dondorp A, Nosten F, Stepniewska K et al (2005a) Artesunate versus quinine for treatment of severe falciparum malaria: a randomised trial. Lancet 366:717–725

Dondorp AM, Desakorn V, Pongtavornpinyo W et al (2005b) Estimation of the total parasite biomass in acute falciparum malaria from plasma PfHRP2. PLoS Med 2:e204

Dondorp AM, Fanello CI, Hendriksen IC et al (2010) Artesunate versus quinine in the treatment of severe falciparum malaria in African children (AQUAMAT): an open-label, randomised trial. Lancet 376:1647–1657

Eckstein-Ludwig U, Webb RJ, Van Goethem ID et al (2003) Artemisinins target the SERCA of Plasmodium falciparum. Nature 424:957–961

Egan TJ (2008) Haemozoin formation. Mol Biochem Parasitol 157:127–136

Epstein JE, Tewari K, Lyke KE et al (2011) Live attenuated malaria vaccine designed to protect through hepatic CD8^{+} T cell immunity. Science 334:475–480

Fitch CD, Cai GZ, Chen YF et al (2003) Relationship of chloroquine-induced redistribution of a neutral aminopeptidase to hemoglobin accumulation in malaria parasites. Arch Biochem Biophys 410:296–306

Fowkes FJ, Richards JS, Simpson JA et al (2010) The relationship between anti-merozoite antibodies and incidence of Plasmodium falciparum malaria: a systematic review and meta-analysis. PLoS Med 7:e1000218

Gamboa D, Ho MF, Bendezu J et al (2010) A large proportion of P. falciparum isolates in the Amazon region of Peru lack pfhrp2 and pfhrp3: implications for malaria rapid diagnostic tests. PLoS One 5:e8091

Ghosh AK, Ribolla PE, Jacobs-Lorena M (2001) Targeting Plasmodium ligands on mosquito salivary glands and midgut with a phage display peptide library. Proc Natl Acad Sci U S A 98:13278–13281

Goodman AL, Draper SJ (2010) Blood-stage malaria vaccines—recent progress and future challenges. Ann Trop Med Parasitol 104:189–211

Gordon DM, Mcgovern TW, Krzych U et al (1995) Safety, immunogenicity, and efficacy of a recombinantly produced Plasmodium falciparum circumsporozoite protein-hepatitis B surface antigen subunit vaccine. J Infect Dis 171:1576–1585

Greenwood B (2005) Malaria vaccines. Evaluation and implementation. Acta Trop 95:298–304

Greenwood BM, Fidock DA, Kyle DE et al (2008) Malaria: progress, perils, and prospects for eradication. J Clin Invest 118:1266–1276

Grobusch MP, Egan A, Gosling RD et al (2007) Intermittent preventive therapy for malaria: progress and future directions. Curr Opin Infect Dis 20:613–620

Gwadz RW, Kaslow D, Lee JY et al (1989) Effects of magainins and cecropins on the sporogonic development of malaria parasites in mosquitoes. Infect Immun 57:2628–2633

Han ET, Watanabe R, Sattabongkot J et al (2007) Detection of four Plasmodium species by genus- and species-specific loop-mediated isothermal amplification for clinical diagnosis. J Clin Microbiol 45:2521–2528

Haynes RK (2006) From artemisinin to new artemisinin antimalarials: biosynthesis, extraction, old and new derivatives, stereochemistry and medicinal chemistry requirements. Curr Top Med Chem 6:509–537

Haynes RK, Chan WC, Wong HN et al (2010) Facile oxidation of leucomethylene blue and dihydroflavins by artemisinins: relationship with flavoenzyme function and antimalarial mechanism of action. ChemMedChem 5:1282–1299

Haynes RK, Cheu KW, Chan HW et al (2012) Interactions between artemisinins and other antimalarial drugs in relation to the cofactor model—a unifying proposal for drug action. ChemMedChem 7:2204–2226

Higgins SJ, Xing K, Kim H et al (2013) Systemic release of high mobility group box 1 (HMGB1) protein is associated with severe and fatal Plasmodium falciparum malaria. Malar J 12:105

Hoppe HC, Van Schalkwyk DA, Wiehart UIM et al (2004) Antimalarial quinolines and artemisinin inhibit endocytosis in Plasmodium falciparum. Antimicrob Agents Chemother 48:2370–2378

Houzé S, Hubert V, Le Pessec G et al (2011) Combined deletions of pfhrp2 and pfhrp3 genes result in Plasmodium falciparum malaria false-negative rapid diagnostic test. J Clin Microbiol 49:2694–2696

Hurwitz I, Fieck A, Read A et al (2011) Paratransgenic control of vector borne diseases. Int J Biol Sci 7:1334–1344

Hviid L (2010) The role of Plasmodium falciparum variant surface antigens in protective immunity and vaccine development. Hum Vaccin 6:84–89

Iqbal J, Khalid N, Hira PR (2002) Comparison of two commercial assays with expert microscopy for confirmation of symptomatically diagnosed malaria. J Clin Microbiol 40:4675–4678

Iseki H, Kawai S, Takahashi N et al (2010) Evaluation of a loop-mediated isothermal amplification method as a tool for diagnosis of infection by the zoonotic simian malaria parasite Plasmodium knowlesi. J Clin Microbiol 48:2509–2514

Jani D, Nagarkatti R, Beatty W et al (2008) HDP-a novel heme detoxification protein from the malaria parasite. PLoS Pathog 4:e1000053

Jepsen MP, Jogdand PS, Singh SK et al (2013) The malaria vaccine candidate GMZ2 elicits functional antibodies in individuals from malaria endemic and non-endemic areas. J Infect Dis 208:479–488

Kattenberg JH, Versteeg I, Migchelsen SJ et al (2012) New developments in malaria diagnostics: monoclonal antibodies against Plasmodium dihydrofolate reductase-thymidylate synthase, heme detoxification protein and glutamate rich protein. MAbs 4:120–126

Khadjavi AG, Prato M (2010) From control to eradication of malaria: the end of being stuck in second gear? Asian Pac J Trop Med 3:412–420

Kilian AH, Metzger WG, Mutschelknauss EJ et al (2000) Reliability of malaria microscopy in epidemiological studies: results of quality control. Trop Med Int Health 5:3–8
Kokoza V, Ahmed A, Woon Shin S et al (2010) Blocking of Plasmodium transmission by cooperative action of Cecropin A and Defensin A in transgenic Aedes aegypti mosquitoes. Proc Natl Acad Sci U S A 107:8111–8116
Krungkrai J, Burat D, Kudan S et al (1999) Mitochondrial oxygen consumption in asexual and sexual blood stages of the human malarial parasite, Plasmodium falciparum. Southeast Asian J Trop Med Public Health 30:636–642
Kumar N, Pande V, Bhatt RM et al (2013) Genetic deletion of HRP2 and HRP3 in Indian Plasmodium falciparum population and false negative malaria rapid diagnostic test. Acta Trop 125:119–121
Kutz FW, Wood PH, Bottimore DP (1991) Organochlorine pesticides and polychlorinated biphenyls in human adipose tissue. Rev Environ Contam Toxicol 120:1–82
Lau YL, Fong MY, Mahmud R et al (2011) Specific, sensitive and rapid detection of human plasmodium knowlesi infection by loop-mediated isothermal amplification (LAMP) in blood samples. Malar J 10:197
Legrand E, Volney B, Meynard JB et al (2008) In vitro monitoring of Plasmodium falciparum drug resistance in French Guiana: a synopsis of continuous assessment from 1994 to 2005. Antimicrob Agents Chemother 52:288–298
Longo M, Zanoncelli S, Torre PD et al (2006) In vivo and in vitro investigations of the effects of the antimalarial drug dihydroartemisinin (DHA) on rat embryos. Reprod Toxicol 22:797–810
Mayxay M, Pukrittayakamee S, Chotivanich K et al (2001) Persistence of Plasmodium falciparum HRP-2 in successfully treated acute falciparum malaria. Trans R Soc Trop Med Hyg 95:179–182
Mayxay M, Pukrittayakamee S, Newton PN et al (2004) Mixed-species malaria infections in humans. Trends Parasitol 20:233–240
Mayxay M, Barends M, Brockman A et al (2007) In vitro antimalarial drug susceptibility and pfcrt mutation among fresh Plasmodium falciparum isolates from the Lao PDR (Laos). Am J Trop Med Hyg 76:245–250
Meshnick SR, Thomas A, Ranz A et al (1991) Artemisinin (qinghaosu): the role of intracellular hemin in its mechanism of antimalarial action. Mol Biochem Parasitol 49:181–189
Mharakurwa S, Shiff CJ (1997) Post treatment sensitivity studies with the ParaSight-F test for malaria diagnosis in Zimbabwe. Acta Trop 66:61–67
Minota S, Cameron B, Welch WJ et al (1988) Autoantibodies to the constitutive 73-kD member of the hsp70 family of heat shock proteins in systemic lupus erythematosus. J Exp Med 168:1475–1480
Mouatcho JC, Goldring JP (2013) Malaria rapid diagnostic tests: challenges and prospects. J Med Microbiol 62:1491–1505
Mueller I, Betuela I, Ginny M et al (2007) The sensitivity of the OptiMAL rapid diagnostic test to the presence of Plasmodium falciparum gametocytes compromises its ability to monitor treatment outcomes in an area of Papua New Guinea in which malaria is endemic. J Clin Microbiol 45:627–630
Müller P, Warr E, Stevenson BJ et al (2008) Field-caught permethrin-resistant Anopheles gambiae overexpress CYP6P3, a P450 that metabolises pyrethroids. PLoS Genet 4:e1000286
Mutanda LN (1999) Assessment of drug resistance to the malaria parasite in residents of Kampala, Uganda. East Afr Med J 76:421–424
N'Guessan R, Darriet F, Guillet P et al (2003) Resistance to carbosulfan in Anopheles gambiae from Ivory Coast, based on reduced sensitivity of acetylcholinesterase. Med Vet Entomol 17:19–25
Na BK, Park JW, Lee HW et al (2007) Characterization of Plasmodium vivax heat shock protein 70 and evaluation of its value for serodiagnosis of tertian malaria. Clin Vaccine Immunol 14:320–322
Nardin EH, Nussenzweig RS (1993) T cell responses to pre-erythrocytic stages of malaria: role in protection and vaccine development against pre-erythrocytic stages. Annu Rev Immunol 11:687–727

Nguyen PH, Day N, Pram TD et al (1995) Intraleucocytic malaria pigment and prognosis in severe malaria. Trans R Soc Trop Med Hyg 89:200–204

Nussenzweig RS, Vanderberg J, Most H et al (1967) Protective immunity produced by the injection of x-irradiated sporozoites of plasmodium berghei. Nature 216:160–162

O'Neill PM, Bray PG, Hawley SR et al (1998) 4-Aminoquinolines—past, present, and future: a chemical perspective. Pharmacol Ther 77:29–58

O'Neill PM, Ward SA, Berry NG et al (2006) A medicinal chemistry perspective on 4-aminoquinoline antimalarial drugs. Curr Top Med Chem 6:479–507

O'Neill PM, Barton VE, Ward SA (2010) The molecular mechanism of action of artemisinin—the debate continues. Molecules 15:1705–1721

Oddoux O, Debourgogne A, Kantele A et al (2011) Identification of the five human Plasmodium species including P. knowlesi by real-time polymerase chain reaction. Eur J Clin Microbiol Infect Dis 30:597–601

Olliaro P (2001) Mode of action and mechanisms of resistance for antimalarial drugs. Pharmacol Ther 89:207–219

Olliaro PL, Haynes RK, Meunier B et al (2001) Possible modes of action of the artemisinin-type compounds. Trends Parasitol 17:122–126

Palmer CJ, Lindo JF, Klaskala WI et al (1998) Evaluation of the OptiMAL test for rapid diagnosis of Plasmodium vivax and Plasmodium falciparum malaria. J Clin Microbiol 36:203–206

Partnership RBM (2008) The Global Malaria Action Plan—for a malaria free world.

Peters W (1982) Antimalarial drug resistance: an increasing problem. Br Med Bull 38:187–192

Poon LL, Wong BW, Ma EH et al (2006) Sensitive and inexpensive molecular test for falciparum malaria: detecting Plasmodium falciparum DNA directly from heat-treated blood by loop-mediated isothermal amplification. Clin Chem 52:303–306

Posner GH, Oh CH, Gerena L et al (1992) Extraordinarily potent antimalarial compounds: new, structurally simple, easily synthesized, tricyclic 1,2,4-trioxanes. J Med Chem 35:2459–2467

Pradel G (2007) Proteins of the malaria parasite sexual stages: expression, function and potential for transmission blocking strategies. Parasitology 134:1911–1929

Pradines B, Mabika Mamfoumbi M, Parzy D et al (1998) In vitro susceptibility of Gabonese wild isolates of Plasmodium falciparum to artemether, and comparison with chloroquine, quinine, halofantrine and amodiaquine. Parasitology 117(Pt 6):541–545

Price RN, Douglas NM, Anstey NM (2009) New developments in Plasmodium vivax malaria: severe disease and the rise of chloroquine resistance. Curr Opin Infect Dis 22:430–435

Ranson H, N'Guessan R, Lines J et al (2011) Pyrethroid resistance in African anopheline mosquitoes: what are the implications for malaria control? Trends Parasitol 27:91–98

Raynes K (1999) Bisquinoline antimalarials: their role in malaria chemotherapy. Int J Parasitol 29:367–379

Richards JS, Beeson JG (2009) The future for blood-stage vaccines against malaria. Immunol Cell Biol 87:377–390

Ridley RG, Dorn A, Vippagunta SR et al (1997) Haematin (haem) polymerization and its inhibition by quinoline antimalarials. Ann Trop Med Parasitol 91:559–566

Rieckmann KH, Carson PE, Beaudoin RL et al (1974) Letter: sporozoite induced immunity in man against an Ethiopian strain of Plasmodium falciparum. Trans R Soc Trop Med Hyg 68: 258–259

Roberts L, Egan TJ, Joiner KA et al (2008) Differential effects of quinoline antimalarials on endocytosis in Plasmodium falciparum. Antimicrob Agents Chemother 52:1840–1842

Rogerson SJ, Hviid L, Duffy PE et al (2007) Malaria in pregnancy: pathogenesis and immunity. Lancet Infect Dis 7:105–117

Sattabongkot J, Tsuboi T, Hisaeda H et al (2003) Blocking of transmission to mosquitoes by antibody to Plasmodium vivax malaria vaccine candidates Pvs25 and Pvs28 despite antigenic polymorphism in field isolates. Am J Trop Med Hyg 69:536–541

Seder RA, Chang LJ, Enama ME et al (2013) Protection against malaria by intravenous immunization with a nonreplicating sporozoite vaccine. Science 341:1359–1365

Sharma YD (1992) Structure and possible function of heat-shock proteins in Falciparum malaria. Comp Biochem Physiol B 102:437–444
Sherman IW (1979) Biochemistry of Plasmodium (malarial parasites). Microbiol Rev 43:453–495
Sinden RE (2010) A biologist's perspective on malaria vaccine development. Hum Vaccin 6:3–11
Snounou G, Viriyakosol S, Jarra W et al (1993) Identification of the four human malaria parasite species in field samples by the polymerase chain reaction and detection of a high prevalence of mixed infections. Mol Biochem Parasitol 58:283–292
Srivastava IK, Morrisey JM, Darrouzet E et al (1999) Resistance mutations reveal the atovaquone-binding domain of cytochrome b in malaria parasites. Mol Microbiol 33:704–711
Sutherland CJ (2009) Surface antigens of Plasmodium falciparum gametocytes—a new class of transmission-blocking vaccine targets? Mol Biochem Parasitol 166:93–98
Tachibana M, Sato C, Otsuki H et al (2012) Plasmodium vivax gametocyte protein Pvs230 is a transmission-blocking vaccine candidate. Vaccine 30:1807–1812
Thévenon AD, Zhou JA, Megnekou R et al (2010) Elevated levels of soluble TNF receptors 1 and 2 correlate with Plasmodium falciparum parasitemia in pregnant women: potential markers for malaria-associated inflammation. J Immunol 185:7115–7122
Tinto H, Rwagacondo C, Karema C et al (2006) In-vitro susceptibility of Plasmodium falciparum to monodesethylamodiaquine, dihydroartemisinin and quinine in an area of high chloroquine resistance in Rwanda. Trans R Soc Trop Med Hyg 100:509–514
Touré AO, Koné LP, Jambou R et al (2008) In vitro susceptibility of P. falciparum isolates from Abidjan (Côte d'Ivoire) to quinine, artesunate and chloroquine. Sante 18:43–47
Vaidya AB, Lashgari MS, Pologe LG et al (1993) Structural features of Plasmodium cytochrome b that may underlie susceptibility to 8-aminoquinolines and hydroxynaphthoquinones. Mol Biochem Parasitol 58:33–42
Vanbuskirk KM, O'Neill MT, De La Vega P et al (2009) Preerythrocytic, live-attenuated Plasmodium falciparum vaccine candidates by design. Proc Natl Acad Sci U S A 106: 13004–13009
Vennerstrom JL, Ellis WY, Ager AL et al (1992) Bisquinolines. 1. N, N-bis(7-chloroquinolin-4-yl) alkanediamines with potential against chloroquine-resistant malaria. J Med Chem 35:2129–2134
Vennerstrom JL, Nuzum EO, Miller RE et al (1999) 8-Aminoquinolines active against blood stage Plasmodium falciparum in vitro inhibit hematin polymerization. Antimicrob Agents Chemother 43:598–602
Vinayak S, Rathore D, Kariuki S et al (2009) Limited genetic variation in the Plasmodium falciparum heme detoxification protein (HDP). Infect Genet Evol 9:286–289
Warhurst DC, Williams JE (1996) ACP Broadsheet no 148. July 1996. Laboratory diagnosis of malaria. J Clin Pathol 49:533–538
Wellems TE, Plowe CV (2001) Chloroquine-resistant malaria. J Infect Dis 184:770–776
White NJ, Pukrittayakamee S, Hien TT et al (2014) Malaria. Lancet 383:723–735
WHO (2010) Guidelines for the treatment of malaria. WHO, Geneva
WHO (2011) Malaria rapid diagnostic test performance: summary results of WHO malaria RDT product testing: rounds 1–3 (2008–2011)
WHO (2012) Global plan for insecticide resistance management in malaria vectors. WHO, Geneva
WHO (2013) World malaria report. WHO, Geneva
Wilson ML (2012) Malaria rapid diagnostic tests. Clin Infect Dis 54:1637–1641
Wu Y, Ellis RD, Shaffer D et al (2008) Phase 1 trial of malaria transmission blocking vaccine candidates Pfs25 and Pvs25 formulated with montanide ISA 51. PLoS One 3:e2636
Yuvaniyama J, Chitnumsub P, Kamchonwongpaisan S et al (2003) Insights into antifolate resistance from malarial DHFR-TS structures. Nat Struct Biol 10:357–365
Zhang M, Hisaeda H, Kano S et al (2001) Antibodies specific for heat shock proteins in human and murine malaria. Microbes Infect 3:363–367

Chapter 3
Lysozymes in the Animal Kingdom

Vivian Tullio, Roberta Spaccapelo, and Manuela Polimeni

1 Introduction

"Lysozyme, which was described by Fleming, in 1922, is an enzyme which is able to destroy not only the non-pathogenic bacteria, but also most of the organisms pathogenic for man and the lower animals. It is very widely distributed in Nature, being found in all animal tissues and tissue fluids and in most of the secretions. It is destructive for many of the pathogenic bacteria in the tears, polymorphonuclear leucocytes, nasal mucus and sputum of man. This being so, it must be of the greatest importance in the defense of the normal individual against bacterial invasion." These words were written by Frederick Thomas Ridley (1904–1977) in June 1928 (Ridley 1928) (see Fig. 3.1). Ridley was born and educated in Birmingham. He graduated in 1922 and became a pathologist at the Central London Ophthalmic Hospital. He worked with Alexander Fleming on lysozyme and devised a system of purification and concentration. He then worked at the Western, in 1928 gained his FRCS, and joined the staff at Moorfields, High Holborn, where he developed micro-corneal lenses for therapeutic use including indolent corneal ulcers, dry eyes, and symblepharon. His rule was never produced on a commercial basis but the prototype is now in the Contact Lens Collection of the BOA Museum at the College of Optometrists. Ridley was President of the ophthalmological section of Royal Society of Medicine in 1963 and retired as Director of the Moorfields contact lens clinic in 1968.

V. Tullio (✉)
Dipartimento di Scienze della Sanità Pubblica e Pediatriche, Università degli Studi di Torino, Torino, Italy
e-mail: vivian.tullio@unito.it

R. Spaccapelo
Dipartimento di Medicina Sperimentale, Università degli studi di Perugia, Perugia, Italy

M. Polimeni
Dipartimento di Oncologia, Università degli Studi di Torino, Torino, Italy

M. Prato (ed.), *Human and Mosquito Lysozymes*,
DOI 10.1007/978-3-319-09432-8_3

Fig. 3.1 A picture of Frederick Thomas Ridley (1904–1977)

Fig. 3.2 Some pictures of Alexander Fleming (1881–1955). (**a**) *Staphylococcus aureus* culture; (**b**) *Penicillium* spp. colonies on agar plates

Sir Alexander Fleming was born at Lochfield near Darvel in Ayrshire, Scotland, on August 6, 1881. He attended the polytechnic and in 1906 began research at St. Mary's under Sir Almroth Wright, a pioneer in vaccine therapy. He gained M.B., B.S. (London), with Gold Medal in 1908, was elected Professor of the School in 1928, and Emeritus Professor of Bacteriology, University of London, in 1948. He was elected Fellow of the Royal Society in 1943 and knighted in 1944. He was Rector of Edinburgh University during 1951–1954. He was also awarded doctorate, honoris causa, degrees of almost 30 European and American Universities. Fleming owes his fame to the discovery of penicillin in 1928, but already in 1921, he observed that a drop of nasal mucus, which accidently fell onto an agar plate as he was suffering from a cold, caused lysis of the bacteria on the plate (see Fig. 3.2). From this observation, he discovered an important bacteriolytic substance, which he later named *Lysozyme* (Fleming 1922). As with many brilliant discoveries, Fleming's skills in looking beyond the obvious led to his discovery of the antimicrobial properties of these two compounds. Sir Alexander wrote numerous papers on bacteriology, immunology, and chemotherapy, including original descriptions of lysozyme and penicillin.

Lysozyme has since been isolated in human tears, saliva, and mother's milk, as well as insects, birds, reptiles, and other mammalian fluids (Newman et al. 1974; McDermott 2013). It is commonly known that lysozymes are distributed in the animal kingdom. Lysozyme is not just present by chance in all these organisms and their related biological fluids: lysozyme is an important part of their immune systems, and plays a key role in the defense of these organisms against bacterial infections. Other types of lysozymes are also found in plants, bacteria, phages, and fungi. In viruses (or bacteriophages), lysozyme is used as an agent to break into the host bacterial cell. Lysozyme from the tail of the virus (or bacteriophage) destroys the peptidoglycan bacterial cell wall and then virus can injects its DNA. After multiplication in bacteria, many lysozyme molecules are created to lyse the bacterial cell wall and release new viruses. Apart from the antimicrobial action, lysozymes also show antifungal and antiviral activities associated with their charge, rather than their lytic activity (Banks et al. 1986; Hung et al. 1999; Losso et al. 2000; Fiołka et al. 2012). These enzymes destroy the AIDS virus (Singh et al. 2005). Lysozymes have many other functions including immune modulation, immune stimulation, and anticancer action (Kovacs-Nolan et al. 2005; Fiołka et al. 2012). In fact, they have important roles in surveillance of membranes of mammalian cells, enhancement of phagocytic activity of macrophages, and stimulation of proliferation and antitumor functions of monocytes (LeMarbre et al. 1981; Bjermer et al. 1986; Sava et al. 1989).

2 Lysozyme Research: Is the Study of Such an Old Molecule Still Worthy Nowadays?

Lysozyme is likely the most scientifically studied protein of all; therefore, it could seem to be an "old" molecule, which is assumed to be known. Lysozyme from chicken egg was first described by Laschtschenko in 1909 (Laschtschenko 1909). Later Bloomfield reported it in saliva in 1919 (Bloomfield and Huck 1920; Imoto 2009). Alexander Fleming officially named lysozyme in 1922, when he discovered a bacterium susceptible to lysozyme, *Micrococcus lysodeikticus*, which is still used today for lysozyme activity assays (Jiang and Huang 2007; Helal and Melzig 2008; Ng et al. 2013). In 1965, Blake et al. solved the structure of lysozyme, making it the second protein and first enzyme structure to be solved by X-ray diffraction methods (Blake et al. 1965). A year later, the mechanism of action of lysozyme was explained (Blake et al. 1967) and nowadays lysozyme is often used as a model in protein biochemistry, enzymology, crystallography, and molecular biology. Throughout the

1960s and in the 1970s, interest in the enzyme increased as a "natural" antibiotic and aid in the diagnosis of disease (Glynn 1968; Pruzanski and Saito 1969). Commercially, the most readily available source of lysozyme has been chicken egg white, from which it is industrially extracted. The lysozyme of chicken egg white has been most extensively studied. Lysozyme has been used in pharmaceutical and food applications for many years, due to its lytic activity on the cell wall of Gram-positive microorganisms. Lysozyme is used in "over-the-counter" drugs in order to increase the natural defenses of the body against bacterial infections. The pharmaceutical use of lysozyme encompasses applications such as oto-rhino-laryngology (lozenges for the treatment of sore throats and of canker sores), and ophthalmology (eye drops and solutions for the decontamination of contact lenses) (McDermott 2013). Lysozyme is also added to infant formulae in order to make them more closely resemble human milk (cow's milk contains very low levels of a lysozyme enzyme). Much research has been done on the use of lysozyme as a preservative in food products, particularly in the Far East and Japan. Several applications have been developed and patented, including the treatment of fresh fruits, vegetables, seafood, meat, tofu, sake, and wine. Lysozyme is a safe product which has been positively evaluated by several regulatory agencies (WHO, FDA, Scientific Food Committee of the EU) and which has a safe record of accomplishment during many years of use in the pharmaceutical and food industries (Proctor and Cunningham 1988; Holzapfel et al. 1995; Mahajan et al. 2010; Abeyrathne et al. 2013; Zheng et al. 2013).

Elevated lysozyme levels were found to be present in the urine and serum of leukemia patients (Osserman and Lawlor 1966) and in the cerebrospinal fluid of patients with a central nervous system tumor (Newman et al. 1974). The enzyme has been also detected in granulocytes, cells of the mononuclear phagocytic system, various exocrine glands, cartilage, and the kidney in which lysozyme is serum derived (Reitamo et al. 1978). Lysozyme research in the 1980s included investigating enzyme intermediates (Acharya and Taniuchi 1982; Desmadril and Yon 1984; Ikeguchi et al. 1986), analyzing the protein structure (Delepierre et al. 1982), and performing binding studies (Perraudin and Prieels 1982; Smith-Gill et al. 1984; Nitta et al. 1988; Liu et al. 2013). In the 1990s, transcription control, silencers, and additional binding sites were investigated (Baniahmad et al. 1991; Madhusudan 1992; Bonifer et al. 1997). Recent studies have focused on obtaining more information about gene regulation of lysozyme both in the hen and in other animals (Shimizu et al. 2005), gaining a better understanding of the secondary structure (Schwinté et al. 2002) and refining its use in biochemical applications (Reischl 2004; Zhu et al. 2007).

Therefore, in this context some questions arise. Is it important to continue to speculate on lysozyme? Is it still relevant to study this important substance nowadays? There is still something to discover on lysozyme? Yes, since many aspects concerning its biological role are not yet understood. Hence, Fleming's prophecy "We shall hear more about lysozyme" has certainly been fulfilled. Lysozyme is an antibacterial agent, but evidence is accumulating which suggests that the enzyme might have other functions as well (Callewaert and Michiels 2010).

3 Classification, Structure, and Function of Lysozymes

Lysozyme hydrolyzes the beta-glycosidic linkage between *N*-acetylmuramic acid and *N*-acetyl glucosamine in the peptidoglycan of bacterial cell walls and can bind polymers of *N*-acetyl glucosamine (Arnheim et al. 1973) (see Fig. 3.3). The mature lysozyme of chicken is composed of 128 amino acids. Amino acid sequences of other avian lysozymes are homologous in sequence and differ in only 4–20 amino acids (Arnheim et al. 1973). The lysozyme gene in chickens is expressed tissue specifically in the oviduct and in macrophages. Although there is only one copy of the lysozyme gene, it is regulated differently in the oviduct and macrophages. Regulation of lysozyme in the oviduct utilizes steroid hormones, while a combination of cis-regulatory elements are used during differentiation in macrophages (Shimizu et al. 2005). Three enhancers, a complex promoter, and a negative regulatory element control transcription (Bonifer et al. 1997; Lefevre et al. 2008).

The structure of lysozyme is consistent under a variety of conditions, making it ideal for crystallography studies. The active site of lysozyme consists of a deep crevice, which divides the protein into two domains linked by an alpha helix. One domain (residues 40–85) consists almost entirely of beta-sheet structure, while the second domain (residues 89–99) is more helical (Strynadka and James 1991).

In the animal kingdom three major different types of lysozymes, differing in amino acids sequences, biochemical and enzymatic properties, have been identified, designated as the C-type (chicken or conventional type), the G-type

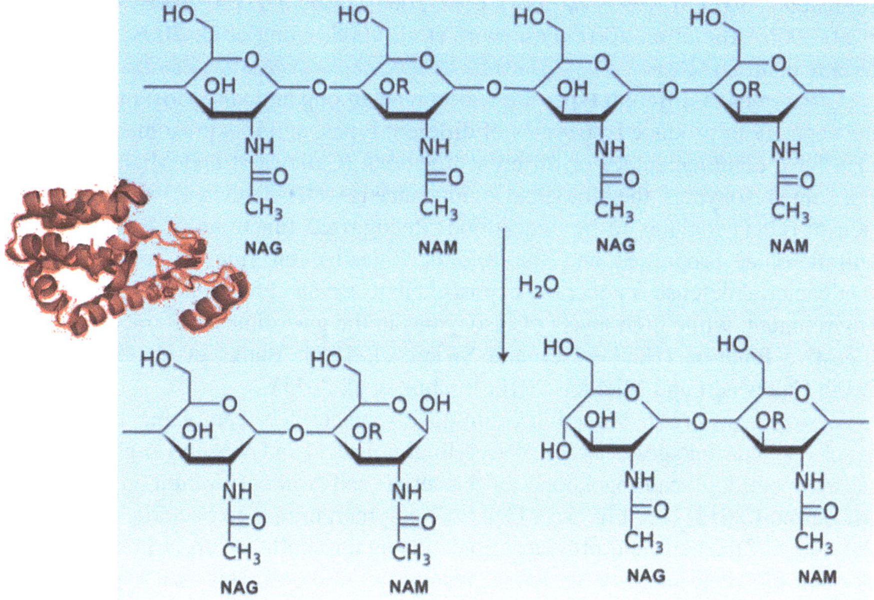

Fig. 3.3 Peptidoglycan and lysozyme catalysis

(goose-type) and, recently, also the I-type (invertebrate type) lysozymes (Callewaert and Michiels 2010).

C-type lysozymes are the archetype molecule, extracted from hen egg white (HEWL). These lysozymes have been served as a model for studies on enzyme structure and function and are the major lysozymes produced by most vertebrates, including mammals (Callewaert and Michiels 2010).

Human lysozyme was the first mammalian lysozyme to be sequenced and, along with HEWL, served as a model protein in many studies (Peters et al. 1989; Prager and Jollès 1996). C-type lysozymes have also been reported in other vertebrates such as birds, fish, reptiles, and amphibians. However, the data for the two latter vertebrate groups are scanty and referred only to turtles and toads (Araki et al. 1998; Thammasirirak et al. 2006; Zhao et al. 2007; Siritapetawee et al. 2009). C-type lysozymes were also found in different classes of Arthropoda, in particular some insect's orders, arachnids (tick and scorpions), and crustaceans (shrimp).

G-type lysozyme was identified in egg whites of the Embden goose (Canfield and McMurry 1967) and then was characterized in chicken, black swan, ostrich, cassowary, and rhea. However, G-type lysozyme is not present only in these birds. In fact, since 2001 many authors have demonstrated lysozyme in several fish species (Hikima et al. 2001; Yin et al. 2003; Sun et al. 2006; Kyomuhendo et al. 2007; Jiménez-Cantizano et al. 2008; Larsen et al. 2009), humans, mice, rats (Irwin and Gong 2003), mollusks (Zhao et al. 2007), and members of the tunicates (Irwin and Gong 2003).

I-type lysozymes belong to the invertebrates: Annelida, Echinodermata, Crustacea, Insects, in particular to mosquitos (*Anopheles gambiae*), Mollusca, and Nematoda (Ito et al. 1999; Zavalova et al. 2004; Takeshita et al. 2004; Matsumoto et al. 2006; Xue et al. 2007; Paskewitz et al. 2008; Cong et al. 2009; Peregrino-Uriarte et al. 2012).

Lysozymes of different types are widespread among animals. Most of them have the capacity to produce lysozymes of different types, and it is presumed that these may have complementary or different functions (Callewaert and Michiels 2010). For some lysozymes, the biological significance is well established, while the function of others remains to be understood. Recognized functions of lysozymes are antimicrobial properties and the role of digestive enzyme in some animals. Antibacterial defense is generally expressed in tissue and body fluids exposed to the environment, while high levels of lysozymes in the gastrointestinal tract indicate a digestive function (Hankiewicz and Swierczek 1974; Banks et al. 1986; Imoto 2009; Callewaert and Michiels 2010; Ibrahim et al. 2011).

Lysozymes were shown to kill Gram-positive bacteria by catalyzing the hydrolysis of 1,4-beta-linkages between *N*-acetylmuramic acid and *N*-acetyl-D-glucosamine in the peptidoglycan backbone of bacterial cell walls (Ibrahim et al. 2011; McDermott 2013) (see Fig. 3.3). Cell wall polymer, unique to bacteria, determines the shape of the cells and provides protection against cellular turgor pressure. Loss of peptidoglycan integrity, therefore, results in rapid cell lysis in a hypo-osmotic environment. On the contrary, the peptidoglycan of Gram-negative bacteria is not directly accessible for lysozymes, because it is surrounded by a lipopolysaccharide-containing outer membrane (Masschalck and Michiels 2003). Nevertheless, this barrier can be breached by components of the innate immune systems of animals such as lactoferrin, defensins, and cathelicidins, which permeabilize the outer membrane, allowing lysozymes to act also against Gram-negative bacteria, including *Pseudomonas aeruginosa* (Itami et al. 1992; Yousif et al. 1994; Hikima et al. 2001; Zheng et al. 2007; Martìnez et al. 2009). In particular, there is evidence that C- and G-type lysozymes isolated from some fishes (i.e., rainbow trout, yellow croacker, and Japanese flounder) have antibacterial activity against Gram-negative pathogen bacteria such as *Vibrio anguillarum*, *V. splendidus*, and *V. parahaemolyticus* (Hikima et al. 1997; Zhao et al. 2007). Moreover, some I-type lysozymes were reported to have significant antibacterial activity against the Gram-negative *Escherichia coli*, *Pediococcus cerevisiae*, and *V. vulnificus*. Recently, it was observed that low concentrations of lysozyme might be beneficial to reduce and prevent biofilm formation by *E. coli* and *Klebsiella pneumoniae subsp. pneumoniae* (Sheffield et al. 2012).

As happened for antibiotics, even in the case of lysozyme bacteria have developed resistance mechanisms. One of these mechanisms is represented by cell wall modifications of the glycan strands such as *N*-deacetylation, *N*-glycolylation, and *O*-acetylation, or the covalent linkage of other cell-wall polymers such as teichoic acid to the peptidoglycan. These cell wall changes have been described in several pathogenic Gram-positive and Gram-negative bacteria and contribute to lysozyme resistance (Zipperle et al. 1984; Clarke and Dupont 1992; Raymond et al. 2005; Bera et al. 2006, 2007; Boneca et al. 2007; Vollmer 2008).

Another bacterial strategy to evade lysozyme action is the production of lysozyme inhibitors (Monchois et al. 2001; Callewaert et al. 2008a, b; Callewaert and Michiels 2010; Wang et al. 2013) that may even contribute to virulence in pathogenic bacteria. For instance the C-type lysozyme inhibitor Ivy is required for the ability of *E. coli* to grow in human saliva, which is naturally rich in lysozyme (Deckers et al. 2008; Callewaert et al. 2009). However, the inhibitory spectrum of the known lysozyme inhibitor families (Ivy family and PliC/MliC family) seems restricted to the C-type lysozymes (Callewaert et al. 2005). Recently, bacterial G-type and I-type lysozyme inhibitors were isolated and identified (Vanderkelen et al. 2008; Van Herreweghe et al. 2010). Further research will have to point out the functional importance of these additional lysozyme inhibitors.

The lysozyme molecule consists of two domains, alfa and beta, between which the active site is located (Ibrahim et al. 2011). A helix-loop-helix peptide at the upper lip of the active site cleft of chicken and human lysozymes confers potent antimicrobial activity with membrane permeabilization action (Ibrahim et al. 2001b).

The molecular mechanism for the antimicrobial function of chicken lysozymes is independent of its catalytic function but appears to depend on a structural phase transition in the molecule (Hankiewicz and Swierczek 1974; Ibrahim 1998; Ibrahim et al. 2001a). In humans, the physiological fluids in which lysozyme exerts its defense role against microorganisms suggest a protease-dependent strategy by which its antimicrobial action is modulated in vivo. In fact, aspartic proteases of azurophil granules, including cathepsin D, have been reported to potentiate the antimicrobial activity of lysozyme against Gram-negative bacteria (Ibrahim et al. 2011). While tears and saliva contain cathepsins G and D, human milk contains cathepsins D, and a significant amount of lysozyme, and seems to play a major role in the protection of infant's gastrointestinal tract. However, aspartic proteases, cathepsins D and E are also involved in severe pathologies of the gastrointestinal mucosa, including cancer and bacterial infections. The functional significance of the coexistence of lysozyme with aspartic proteases on the molecular mechanism of its antimicrobial action seems to be related to a proteolytic process in which a complex of five peptides released by pepsin contributes to the antimicrobial action of lysozyme (Ibrahim et al. 2011). To elucidate this process and the antimicrobial role of the individual peptides of lysozyme assumed to be liberated upon cleavage with neonate pepsin, Ibrahim et al. (2011) synthesized five peptides and tested them for antimicrobial activity against different microorganisms. Experimental findings indicated that there are a variety of peptides within the sequence of lysozyme that are able to contrast Gram-positive and Gram-negative bacteria and the fungus *Candida albicans* by at least two different mechanisms of action: dissipation of bacterial respiration and loss of membrane integrity. Synthetic peptides are able to kill bacteria by crossing the outer membrane of Gram-negative bacteria via self-promoted uptake and are able to dissipate the membrane potential-dependent respiration of Gram-positive bacteria (Ibrahim et al. 2011).

4 Conclusion

Although the first studies on lysozyme date from the 1960s, this enzyme is still a current subject of many investigations and knowledge on lysozyme is still growing. The spread of antibiotic resistance among bacterial pathogen has prompted an effort to explore potential adjuvant and alternative therapies derived from nature's repertoire of bactericidal protein and peptides. Several studies have examined lysozyme's potential as an exogenously administered biotherapeutic by using engineered enzyme that seems an interesting candidate for treating bacterial infection, although further studies will be needed to rigorously assess its therapeutic utility (Bhavsar et al. 2010, 2011; Teneback et al. 2013).

References

Abeyrathne ED, Lee HY, Ahn DU (2013) Egg white proteins and their potential use in food processing or as nutraceutical and pharmaceutical agents—a review. Poult Sci 92:3292–3299

Acharya AS, Taniuchi H (1982) Implication of the structure and stability of disulfide intermediates of lysozyme on the mechanism of renaturation. Mol Cell Biochem 44:129–148

Araki T, Yamamoto T, Torikata T (1998) Reptile lysozyme: the complete amino acid sequence of soft-shelled turtle lysozyme and its activity. Biosci Biotechnol Biochem 62:316–324

Arnheim N, Inouye M, Law L et al (1973) Chemical studies on the enzymatic specificity of goose egg white lysozyme. J Biol Chem 248:233–236
Baniahmad C, Muller M, Altschmied J et al (1991) Co-operative binding of the glucocorticoid receptor DNA binding domain is one of at least two mechanisms for synergism. J Mol Biol 222:155–165
Banks JG, Board RG, Sparks NH (1986) Natural antimicrobial systems and their potential in food preservation of the future. Biotechnol Appl Biochem 8:103–147
Bera A, Biswas R, Herbert S et al (2006) The presence of peptidoglycan O-acetyltransferase in various staphylococcal species correlates with lysozyme resistance and pathogenicity. Infect Immun 74:4598–4604
Bera A, Biswas R, Herbert S et al (2007) Influence of wall teichoic acid on lysozyme resistance in Staphylococcus aureus. J Bacteriol 189:280–283
Bhavsar T, Liu M, Hardej D et al (2010) Aerosolized recombinant human lysozyme ameliorates Pseudomonas aeruginosa induced pneumonia in hamsters. Exp Lung Res 36:94–100
Bhavsar T, Liu M, Liu X et al (2011) Aerosolized recombinant human lysozyme enhances the bactericidal effect of tobramycin in a hamster model of Pseudomonas aeruginosa-induced pneumonia. Exp Lung Res 37:536–541
Bjermer L, Bäck O, Roos G et al (1986) Mast cells and lysozyme positive macrophages in bronchoalveolar lavage from patients with sarcoidosis. Valuable prognostic and activity marking parameters of disease? Acta Med Scand 220:161–166
Blake CC, Johnson LN, Mair GA et al (1967) Crystallographic studies of the activity of hen egg-white lysozyme. Proc R Soc Lond B Biol Sci 167:378–388
Blake CC, Koenig DF, Mair GA et al (1965) Structure of hen egg-white lysozyme. A three-dimensional Fourier synthesis at 2 Angstrom resolution. Nature 206:757–761
Bloomfield AL, Huck JG (1920) The fate of bacteria introduced into the upper air passages: IV. The reaction of the saliva. Bull Johns Hopkins Hosp 31:118–121
Boneca IG, Dussurget O, Cabanes D et al (2007) A critical role for peptidoglycan N-deacetylation in Listeria evasion from the host innate immune system. Proc Natl Acad Sci U S A 104: 997–1002
Bonifer C, Jägle U, Huber MC (1997) The chicken lysozyme locus as a paradigm for the complex developmental regulation of eukaryotic gene loci. J Biol Chem 272:26075–26078
Callewaert L, Aertsen A, Deckers D et al (2008a) A new family of lysozyme inhibitors contributing to lysozyme tolerance in gram-negative bacteria. PLoS Pathog 4:e1000019
Callewaert L, Masschalck B, Deckers D et al (2005) Purification of Ivy, a lysozyme inhibitor from Escherichia coli, and characterisation of its specificity for various lysozymes. Enzyme Microb Technol 37:205–211
Callewaert L, Michiels CW (2010) Lysozymes in the animal kingdom. J Biosci 35:127–160
Callewaert L, Vanderkelen L, Deckers D et al (2008b) Detection of a lysozyme inhibitor in Proteus mirabilis by a new reverse zymogram method. Appl Environ Microbiol 74:4978–4981
Callewaert L, Vanoirbeek KGA, Lurquin I et al (2009) The Rcs two-component system regulates expression of lysozyme inhibitors and is induced by exposure to lysozyme. J Bacteriol 191:1979–1981
Canfield RE, McMurry S (1967) Purification and characterization of a lysozyme from goose egg white. Biochem Biophys Res Commun 26:38–42
Clarke AJ, Dupont C (1992) O-acetylated peptidoglycan: its occurrence, pathobiological significance, and biosynthesis. Can J Microbiol 38:85–91
Cong L, Yang X, Wang X et al (2009) Characterization of an i-type lysozyme gene from the sea cucumber Stichopus japonicus, and enzymatic and nonenzymatic antimicrobial activities of its recombinant protein. J Biosci Bioeng 107:583–588
Deckers D, Vanlint D, Callewaert L et al (2008) Role of Ivy lysozyme inhibitor in growth or survival of Escherichia coli and Pseudomonas aeruginosa in hen egg albumen and in human saliva and breast milk. Appl Environ Microbiol 74:4434–4439
Delepierre M, Dobson CM, Poulsen FM (1982) Studies of β-sheet structure in lysozyme by proton nuclear magnetic resonance. Assignments and analysis of spin-spin coupling constants. Biochemistry 21:4756–4761

Desmadril M, Yon JM (1984) Evidence for intermediates during unfolding and refolding of a two-domain protein, phage T4 lysozyme: equilibrium and kinetic studies. Biochemistry 23:11–19

Fiołka MJ, Zagaja MP, Hułas-Stasiak M et al (2012) Activity and immunodetection of lysozyme in earthworm Dendrobaena veneta (Annelida). J Invertebr Pathol 109:83–90

Fleming A (1922) On a remarkable bacteriolytic element found in tissues and secretions. Proc R Soc Lond Ser B Biol Sci 93:306–317

Glynn AA (1968) Lysozyme: antigen, enzyme and antibacterial agent. Sci Basis Med Annu Rev 31–52

Hankiewicz J, Swierczek E (1974) Lysozyme in human body fluids. Clin Chim Acta 57:205–209

Helal R, Melzig MF (2008) Determination of lysozyme activity by a fluorescence technique in comparison with the classical turbidity assay. Pharmazie 63:415–419

Hikima J, Hirono I, Aoki T (1997) Characterization and expression of c-type lysozyme cDNA from Japanese flounder (Paralichthys olivaceus). Mol Mar Biol Biotechnol 6:339–344

Hikima JI, Minagawa S, Hirono I et al (2001) Molecular cloning, expression and evolution of the Japanese flounder goose-type lysozyme gene, and the lytic activity of its recombinant protein. Biochim Biophys Acta 1520:35–44

Holzapfel WH, Geisen R, Schillinger U (1995) Biological preservation of foods with reference to protective cultures, bacteriocins and food-grade enzymes. Int J Food Microbiol 24:343–362

Hung SC, Wang W, Chan SI et al (1999) Membrane lysis by the antibacterial peptides cecropins B1 and B3: a spin-label electron spin resonance study on phospholipid bilayers. Biophys J 77:3120–3133

Ibrahim HR (1998) On the novel catalytically-independent antimicrobial function of hen egg-white lysozyme: a conformation-dependent activity. Nahrung 42:187–193

Ibrahim HR, Imazato K, Ono H (2011) Human lysozyme possesses novel antimicrobial peptides within its N-terminal domain that target bacterial respiration. J Agric Food Chem 59: 10336–10345

Ibrahim HR, Matsuzaki T, Aoki T (2001a) Genetic evidence that antibacterial activity of lysozyme is independent of its catalytic function. FEBS Lett 506:27–32

Ibrahim HR, Thomas U, Pellegrini A (2001b) A helix-loop-helix peptide at the upper lip of the active site cleft of lysozyme confers potent antimicrobial activity with membrane permeabilization action. J Biol Chem 276:43767–43774

Ikeguchi M, Kuwajima K, Mitani M et al (1986) Evidence for identity between the equilibrium unfolding intermediate and a transient folding intermediate: a comparative study of the folding reactions of α-lactalbumin and lysozyme. Biochemistry 25:6965–6972

Imoto T (2009) Lysozyme. eLS. Published on-line in Wiley Online Library

Irwin DM, Gong Z (2003) Molecular evolution of vertebrate goose-type lysozyme genes. J Mol Evol 56:234–242

Itami T, Takehara A, Nagano Y et al (1992) Purification and characterization from Ayu skin mucus. Nippon Suisan Gakkaishi 58:1937–1944

Ito Y, Yoshikawa A, Hotani T et al (1999) Amino acid sequences of lysozymes newly purified from invertebrates imply wide distribution of a novel class in the lysozyme family. Eur J Biochem 259:456–461

Jiang ZL, Huang GX (2007) Resonance scattering spectra of Micrococcus lysodeikticus and its application to assay of lysozyme activity. Clin Chim Acta 376:136–141

Jiménez-Cantizano RM, Infante C, Martin-Antonio B et al (2008) Molecular characterization, phylogeny, and expression of c-type and g-type lysozymes in brill (Scophthalmus rhombus). Fish Shellfish Immunol 25:57–65

Kovacs-Nolan J, Phillips M, Mine Y (2005) Advances in the value of eggs and egg components for human health. J Agric Food Chem 53:8421–8431

Kyomuhendo P, Myrnes B, Nilsen IW (2007) A cold-active salmon goose-type lysozyme with high heat tolerance. Cell Mol Life Sci 64:2841–2847

Larsen AN, Solstad T, Svineng G et al (2009) Molecular characterisation of a goose-type lysozyme gene in Atlantic cod (Gadus morhua L.). Fish Shellfish Immunol 26:122–132

Laschtschenko P (1909) Über die keimtötende und entwicklungshemmende Wirkung Hühnereiweiß. Z Hyg InfektKrankh (in German) 64:419–427

Lefevre P, Witham J, Lacroix CE et al (2008) The LPS-induced transcriptional upregulation of the chicken lysozyme locus involves CTCF eviction and noncoding RNA transcription. Mol Cell 32:129–139

LeMarbre P, Rinehart JJ, Kay NE et al (1981) Lysozyme enhances monocyte-mediated tumoricidal activity: a potential amplifying mechanism of tumor killing. Blood 58:994–999

Liu J, Luo K, Song Z (2013) Binding study of lysozyme with Al(III) using chemiluminescence analysis. Luminescence. Published on-line in Wiley Online Library

Losso JN, Nakai S, Charter EA (2000) Lysozyme. In: Naidi AS (ed) Natural food antimicrobial system. CRC, New York, pp 185–210

Madhusudan VM (1992) Additional binding sites in lysozyme. X-ray analysis of lysozyme complexes with bromophenol red and bromophenol blue. Protein Eng 5:399–404

Mahajan R, Gupta VK, Sharma J (2010) Comparison and suitability of gel matrix for entrapping higher content of enzymes for commercial applications. Indian J Pharm Sci 72:223–228

Martìnez JG, Waldon M, Huang Q et al (2009) Membrane-targeted synergistic activity of docosahexaenoic acid and lysozyme against Pseudomonas aeruginosa. Biochem J 419:193–200

Masschalck B, Michiels CW (2003) Antimicrobial properties of lysozyme in relation to foodborne vegetative bacteria. Crit Rev Microbiol 29:191–214

Matsumoto T, Nakamura AM, Takahashi KG (2006) Cloning of cDNAs and hybridization analysis of lysozymes from two oyster species, Crassostrea gigas and Ostrea edulis. Comp Biochem Physiol B Biochem Mol Biol 145:325–330

McDermott AM (2013) Antimicrobial compounds in tears. Exp Eye Res 117:53–61

Monchois V, Abergel C, Sturgis J et al (2001) Escherichia coli ykfE ORFan gene encodes a potent inhibitor of c-type lysozyme. J Biol Chem 276:18437–18441

Newman J, Josephson AS, Cacatian A et al (1974) Spinal-fluid lysozyme in the diagnosis of central-nervous-system tumours. Lancet 2:756–757

Ng A, Heynen M, Luensmann D et al (2013) Optimization of a fluorescence-based lysozyme activity assay for contact lens studies. Curr Eye Res 38:252–259

Nitta K, Tsuge H, Shimazaki K et al (1988) Calcium-binding lysozymes. Biol Chem Hoppe Seyler 369:671–675

Osserman EF, Lawlor DP (1966) Serum and urinary lysozyme (muramidase) in monocytic and monomyelocytic leukemia. J Exp Med 124:921–952

Paskewitz SM, Li B, Kajla KM (2008) Cloning and molecular characterization of two invertebrate-type lysozymes from Anopheles gambiae. Insect Mol Biol 17:217–225

Peregrino-Uriarte AB, Muhlia-Almazan AT, Arvizu-Flores AA et al (2012) Shrimp invertebrate lysozyme i-lyz: gene structure, molecular model and response of c and i lysozymes to lipopolysaccharide (LPS). Fish Shellfish Immunol 32:230–236

Perraudin JP, Prieels JP (1982) Lactoferrin binding to lysozyme-treated Micrococcus luteus. Biochim Biophys Acta 718:42–48

Peters CW, Kruse U, Pollwein R et al (1989) The human lysozyme gene. Eur J Biochem 182: 507–516

Prager EM, Jollès P (1996) Animal lysozymes c and g: an overview. In: Jollès P (ed) Lysozymes: model enzymes in biochemistry and biology. Birkhäuser, Basel, pp 9–31

Proctor VA, Cunningham FE (1988) The chemistry of lysozyme and its use as a food preservative and a pharmaceutical. Crit Rev Food Sci Nutr 26:359–395

Pruzanski W, Saito SG (1969) The diagnostic value of lysozyme (muramidase) estimation in biological fluids. Am J Med Sci 258:405–415

Raymond JB, Mahapatra S, Crick DC et al (2005) Identification of the namH gene, encoding the hydroxylase responsible for the N-glycolylation of the mycobacterial peptidoglycan. J Biol Chem 280:326–333

Reischl U (2004) Purification and immunological characterization of recombinant antigens expressed in the form of insoluble aggregates (inclusion bodies). Methods Mol Med 94: 213–224

Reitamo S, Klockars M, Adinolfi M et al (1978) Human lysozyme (origin and distribution in health and disease). Ric Clin Lab 8:211–231

Ridley F (1928) Lysozyme: an antibacterial body present in great concentration in tears, and its relation to infection of the human eye. Proc R Soc Med 21:1495–1506

Sava G, Ceschia V, Pacor S (1989) Mechanism of the antineoplastic action of lysozyme: evidence for host mediated effects. Anticancer Res 9:1175–1180

Schwintě P, Ball V, Szalontai B et al (2002) Secondary structure of proteins adsorbed onto or embedded in polyelectrolyte multilayers. Biomacromolecules 3:1135–1143

Sheffield CL, Crippen TL, Poole TL et al (2012) Destruction of single-species biofilms of Escherichia coli or Klebsiella pneumoniae subsp. pneumoniae by dextranase, lactoferrin, and lysozyme. Int Microbiol 15:185–189

Shimizu M, Losos JK, Gibbins AM (2005) Analysis of an approach to oviduct-specific expression of modified chicken lysozyme genes. Biochem Cell Biol 83:49–60

Singh M, Chand H, Gupta KC (2005) The studies of density, apparent molar volume, and viscosity of bovine serum albumin, egg albumin, and lysozyme in aqueous and RbI, CsI, and DTAB aqueous solutions at 303.15 K. Chem Biodivers 2:809–824

Siritapetawee J, Thammasirirak S, Robinson RC et al (2009) The 1.9Å X-ray structure of egg-white lysozyme from Taiwanese soft-shelled turtle (Trionyx sinensis Wiegmann) exhibits structural differences from the standard chicken-type lysozyme. J Biochem 145:193–198

Smith-Gill SJ, Rupley JA, Pincus MR et al (1984) Experimental identification of a theoretically predicted "left-sided" binding mode for (GlcNAc) 6 in the active site of lysozyme. Biochemistry 23:993–997

Strynadka NC, James MN (1991) Lysozyme revisited: crystallographic evidence for distortion of an N-acetylmuramic acid residue bound in site D. J Mol Biol 220:401–424

Sun BJ, Wang GL, Xie HX et al (2006) Gene structure of goose-type lysozyme in the mandarin fish Siniperca chuatsi with analysis on the lytic activity of its recombinant in Escherichia coli. Aquaculture 252:106–113

Takeshita K, Hashimoto Y, Thuijhata Y et al (2004) Determination of the complete cDNA sequence, construction of expression systems, and elucidation of fibrinolytic activity for Tapes japonica lysozyme. Protein Expr Purif 36:254–262

Teneback CC, Scanlon TC, Wargo MJ et al (2013) Bioengineered lysozyme reduces bacterial burden and inflammation in a murine model of mucoid Pseudomonas aeruginosa lung infection. Antimicrob Agents Chemother 57:5559–5564

Thammasirirak S, Ponkham P, Preecharram S et al (2006) Purification, characterization and comparison of reptile lysozymes. Comp Biochem Physiol C Toxicol Pharmacol 143:209–217

Van Herreweghe JM, Vanderkelen L, Callewaert L et al (2010) Lysozyme inhibitor conferring bacterial tolerance to invertebrate type lysozyme. Cell Mol Life Sci 67:1177–1188

Vanderkelen L, Callewaert L, Van Herreweghe J et al (2008) Identification of a bacterial inhibitor of g-type lysozyme. Commun Agric Appl Biol Sci 73:227–230

Vollmer W (2008) Structural variation in the glycan strands of bacterial peptidoglycan. FEMS Microbiol Rev 32:287–306

Wang C, Hu YH, Sun BG et al (2013) Edwardsiella tarda Ivy, a lysozyme inhibitor that blocks the lytic effect of lysozyme and facilitates host infection in a manner that is dependent on the conserved cysteine residue. Infect Immun 81:3527–3533

Xue QG, Itoh N, Schey KL et al (2007) A new lysozyme from the eastern oyster (Crassostrea virginica) indicates adaptive evolution of i-type lysozyme. Cell Mol Life Sci 64:82–95

Yin ZX, He JG, Deng WX et al (2003) Molecular cloning, expression of orange spotted grouper goose-type lysozyme cDNA, and lytic activity of its recombinant protein. Dis Aquat Organ 55:117–123

Yousif AN, Albright LJ, Evelyn TPT (1994) In vitro evidence for the antibacterial role of lysozyme in salmonid eggs. Dis Aquat Organ 19:15–19

Zavalova LL, Baskova IP, Barsova EV et al (2004) Recombinant destabilise-lysozyme: synthesis de novo in E. coli and action mechanism of the enzyme expressed in Spodoptera frugiperda. Biochemistry 69:776–781

Zhao J, Song L, Li C et al (2007) Molecular cloning of an invertebrate goose-type lysozyme gene from Chlamys farreri, and lytic activity of the recombinant protein. Mol Immunol 44: 1198–1208

Zheng J, Zhu Y, Yan Q et al (2013) Recombination and functional studies of a dual-action peptide for diabetes. J Drug Target 21:443–449

Zheng W, Tian C, Chen X (2007) Molecular characterization of goose-type lysozyme homologue of large yellow croaker and its involvement in immune response induced by trivalent bacterial vaccine as an acute-phase protein. Immunol Lett 113:107–116

Zhu J, Li D, Jin J et al (2007) Binding analysis of farrerol to lysozyme by spectroscopic methods. Spectrochim Acta A Mol Biomol Spectrosc 68:354–359

Zipperle GF Jr, Ezzell JW Jr, Doyle RJ (1984) Glucosamine substitution and muramidase susceptibility in Bacillus anthracis. Can J Microbiol 30:553–559

Chapter 4
Role of Lysozymes of *Anopheles Mosquitoes* in *Plasmodium* Development

Clelia Oliva*, Luca Facchinelli*, Nicoletta Basilico, and Roberta Spaccapelo

1 Introduction

The hydrolytic enzyme lysozymes are widely found in all living organisms. They are important participants of the antibacterial defense but may also show a digestive function (Callewaert and Michiels 2010). Three major distinct lysozyme types have been identified (c-type, g-type, and i-type) with a common ability to hydrolyze the glycosidic bond between *N*-acetylmuramic acid and *N*-acetyl glucosamine in the peptidoglycan layer of bacterial cell walls.

The antibacterial activity of lysozymes has been demonstrated in most organisms. Bacterial challenge or wounding induces a higher expression of lysozyme genes. The muramidase activity results in the loss of cell wall integrity and the lysis of susceptible bacteria or inhibition of cell growth (Nakimbugave et al. 2006). However, the existence of nonenzymatic bactericidal pathways has been put forward. This may act through the activation of bacterial autolysins or induction of membrane leakage following direct interaction with the cell membrane (During et al. 1996; Ibrahim et al. 2001; Masschalck and Michiels 2003). Lysozyme strongly affects numerous Gram-positive bacteria species and to a lesser extent Gram-negative ones (in insects: Abraham et al. 1995; Yu et al. 2002; Skerrett 2004; Mai and Hu 2009) in which the peptidoglycan layer is shielded by the outer layer of lipopolysaccharide and protein (Masschalck and Michiels 2003). Besides their direct bactericidal activity, lysozymes may be important regulators of the overall

*Author contributed equally with all other contributors.

C. Oliva (✉) • L. Facchinelli • R. Spaccapelo
Dipartimento di Medicina Sperimentale, Università degli Studi di Perugia, Perugia, Italy
e-mail: cleliaoliva@gmail.com

N. Basilico
Dipartimento di Scienze Biomediche, Chirurgiche e Odontoiatriche, Università degli Studi di Milano, Milan, Italy

M. Prato (ed.), *Human and Mosquito Lysozymes*,
DOI 10.1007/978-3-319-09432-8_4

response to bacteria. The interaction of peptidoglycan recognition proteins with the lysozyme-digested peptidoglycans activates the prophenoloxidase cascade leading to melanization (Christensen et al. 2005; Park et al. 2007; Kim et al. 2008).

Insects are the only invertebrates to possess both c- and i-type lysozymes, which suggests that each type might have evolved to fill diverse functional roles (Paskewitz et al. 2008). C-type lysozymes (Hultmark 1996) are the most studied group and have a 35–40 % sequence homology and share a common three-dimensional fold to alpha-lactalbumin. The i-types differ from the c-types in their primary sequence and in electric charge (acidic/neutral vs. basic, respectively) and are shown to have antibacterial activity (Ito et al. 1999; Nilsen et al. 1999; Zavalova et al. 2000; Bachali et al. 2002), although they miss potentially critical amino acids for the muramidase activity (Bachali et al. 2002).

In the mosquito *Anopheles gambiae*, lysozymes are present in different tissues and developmental stage expression profiles (Li et al. 2005; Paskewitz et al. 2008), which probably enhance the response to the bacteria population corresponding to different diet and/or environments. Recently, lysozymes have been shown to play an important role in the development of *Plasmodium* parasite in *Anopheles* species, protecting the oocytes from melanization. Their potential importance for the development of malaria control tools is discussed.

2 Lysozymes in *Anopheles* Mosquito

The first isolation and characterization of a gene encoding a lysozyme (now known as *LysC1*) in *An. gambiae* was reported in 1996 by Kang et al. They showed a strong expression of the transcript in sugar-fed females and low levels of proteins after blood feeding (Kang et al. 1996). It is suggested that lysozyme could be involved in the digestion process of the bacteria and fungi present in the nectar, similarly to *Lys P* in Drosophila (Kylsten et al. 1992). In the higher flies, *Musca domestica* and *Drosophila melanogaster*, lysozymes occurring in the gut exhibit isoelectric points that are adaptive for a digestive function under acidic conditions (Lemos et al. 1993; Daffre et al. 1994). The presence of these enzymes in the salivary glands also suggests a role in the prevention of bacterial infection of the mouthparts (Rossignol and Lueders 1986; Moreira-Ferro et al. 1999).

In *An. gambiae*, eight different lysozymes belonging to the c-type have been discovered (Kajla et al. 2010). Their functional roles are still not completely understood, but they probably possess diverse function and target diverse tissues (Li et al. 2005). *LysC1* and *LysC2* are the most documented proteins and are involved in the innate immunity. The gene expression profiles and the analyses of the predicted proteins suggest that the remaining six genes might be involved in novel functions in immunity or other biological processes. *LysC4*, *LysC5* and *LysC7*, and several of the domains of *LysC6* are lacking critical amino acids for muramidase activity. However, these proteins might still possess an antibacterial activity, which could derive from their ability to bind to *N*-acetyl glucosamine or other oligosaccharides (Li et al. 2005). *LysC4* and *LysC7* transcripts did not increase following bacterial infection or wounding, which makes them unlikely to be involved in immunity. The function of *LysC3* and *LysC8* has not yet been unrevealed, but the presence of a potential calcium binding site suggests that they could be involved in the digestion of bacteria (Li et al. 2005).

I-type lysozymes have been little studied yet and their functional roles in mosquito biology are still unclear. Two genes belonging to the i-type have been discovered in *Anopheles* (Paskewitz et al. 2008). *LysI1* and *LysI2* are expressed in all developmental stages of *An. gambiae* females but not in the salivary glands and in the midgut of non-blood-fed females (Paskewitz et al. 2008). Blood feeding strongly increases the transcript levels of *LysI1* in the ovaries, Malpighian tubules, and fat bodies. *LysI1* and *LysI2* are both upregulated in the mosquito midgut after blood feeding. The expression of i-type lysozymes in the gut could suggest a digestive rather than an immune function. The involvement of these proteins in the immunity has not yet been demonstrated, and wounding or injection of *Micrococcus luteus* did not affect the transcription of *LysI1* in *An. gambiae* but consistently downregulates *LysI2* transcripts (Paskewitz et al. 2008). These enzymes could be involved in the digestion of bacteria present in the blood or in the breaking down of the blood clots (Zavalova et al. 2000; Paskewitz et al. 2008).

3 Lysozyme C1 and *Anopheles* Immune System

Kajla et al. (2010) stated that lysozyme C1 is constitutively expressed in the midgut and in the salivary glands of *An. gambiae* but the same researchers failed to detect it in the midgut in a later study (Kajla et al. 2011). Bacterial challenge upregulates the expression of *LysC1* gene at least up to 72 h posttreatment, induces a strong increase of the protein in the hemolymph and a higher muramidase activity from 15 to 120 h posttreatment (Li et al. 2005; Dong et al. 2006, 2009; Kajla et al. 2010). However LysC1 directly kills only a few bacteria species but seems to play an important indirect role in the immune response. Indeed, the knocking down of the gene increased the mosquito mortality after infection with the Gram-negative *E. coli* although the bacteria were not killed in vitro by the enzyme (Kajla et al. 2010). Kajla et al. (2010) showed that the knocking down of *LysC1* does not affect the transcription of other genes involved in the immune response. It is therefore hypothesized that the production of small peptidoglycan fragments by *LysC*1 might upregulate the signaling cascades that result in the production of antimicrobial peptides.

4 Interaction of Lysozyme and *Plasmodium*

The sporogonic development of malaria parasites depends on a complex interaction with their mosquito hosts. In *An. gambiae*, LysC1 binds to and can protect an abiotic target (CM-Sephadex beads) from melanization (Li and Paskewitz 2006).

Kajla et al. (2011) showed through immunohistochemical analyses and gene silencing that physical interaction of LysC1 with the parasite surface following the critical period of midgut invasion was associated with parasite persistence. The injection of dsRNA into the thorax of female *An. gambiae* G3 mosquitoes significantly reduced the expression of *LysC*1. Four days after dsRNA injection, mosquitoes were allowed to feed on mice infected with GFP-expressing *Plasmodium berghei.* Three days post-infection the number of oocysts per midgut were scored showing that knockdown of *LysC*1 significantly reduced prevalence and intensity of *P. berghei* infections (Kajla et al. 2011). Similar results were obtained in a different study where the knockdown of *AdLys C1* gene in *Anopheles dirus* showed the agonistic role of *LysC1* in the response of mosquitoes during *P. berghei* infection (Kajla et al. 2010).

Knockdown of *LysC*1 in *An. gambiae* did not result in changes in numbers of viable *P. berghei* parasites until 3 days post-infection (Kajla et al. 2011). Similar numbers of fluorescing parasites were seen in control and knockdown mosquitoes at 24 h post-infection (Kajla et al. 2011). This suggested that formation of ookinetes and invasion of the midgut were similar in treated and control mosquitoes and that the block occurred after oocysts formation (Kajla et al. 2011). The transition to oocysts occurs once the ookinetes move between the epithelial cells and the midgut basal lamina (BL). The rapidly expanding oocysts stretch the overlying layer of the BL at the hemocelic surface while a new BL is generated between the oocysts and the epithelial cells (Meis et al. 1989). At the same time, mosquito-derived collagen and laminin are incorporated into oocyst capsules (Dessens et al. 2003; Osta et al. 2004; Adini and Warburg 1999; Castillo et al. 2006). Knockdown of laminin mRNA led to a substantial reduction in the number of successfully developed oocysts (Arrighi et al. 2005). Laminin has been shown to bind to at least five *P. berghei* proteins (P25, P28, SOAP, circumsporozoite, and TRAP related) in yeast two hybrid assays (Meis et al. 1989; Dessens et al. 2003; Vlachou et al. 2001). Nacer et al. (2008) showed that mosquito-produced laminin indeed becomes part of the parasite capsule during its passage through the gut. The acquisition of the basal lamina proteins is likely to help protect the developing oocysts from the mosquito immune system and, therefore, may facilitate their prolonged extracellular development in the mosquito body cavity (Castillo et al. 2006). Vertebrate lysozymes bind to glycosaminoglycans in extracellular matrices (Mahairaki et al. 2005) and insect basal laminae are negatively charged (Moss et al. 1997), which could promote interaction with the basic LysC1. Lysozymes have also been shown to bind and prevent the proteolytic degradation of the elastin component of elastic fibers in the basal lamina, indicating that lysozyme interaction can protect elastic fibers at the sites of injury (Park et al. 1996). Arrighi et al. (2005) suggested that the production of new basal lamina around the midgut may be a normal process following blood feeding, a process that has been co-opted by the parasite. Kajla et al. (2011) hypothesize that LysC1 might associate with components of the midgut BL and become incorporated during formation of the BL-related capsule around the parasite. Immunohistochemistry data on the interaction of LysC1 and malaria oocysts support a direct LysC1 association with the parasite (Kajla et al. 2011). Since the detection of LysC1 in Western blots failed and after extended incubation periods of midgut extracts muramidase activity could not be detected, Kajla et al. (2011) speculated that the protein may not originate from the midgut cells. Ahmed et al. (2002) also failed to detect muramidase activity in midgut extracts following blood feeding. By contrast Kajla et al. (2011) detected LysC1 in mosquito hemolymph through Western blotting (Li and Paskewitz 2006; Kajla et al. 2010) and Ahmed et al. (2002) determined that muramidase activity in the hemolymph increased following blood feeding. Castillo et al. (2006) also described the occurrence of LysC1 in hemocytes. Together, these observations suggest that LysC1 associated with parasites is derived from the hemolymph. In studies of the transport of molecules from the hemolymph across the basal lamina to the intercellular spaces of the midgut epithelium, other researchers have shown that cytochrome-*c* can make this passage (Reddy and Locke 1990). Cytochrome-*c* is nearly identical to LysC1 in both size and charge. Thus, it seems likely that LysC1 can also move in this direction.

Rao et al. (2010) suggested that the trade-off between lysozyme activity and phenoloxydase activity (PO) (Cotter et al. 2008; Povey et al. 2009) might result in the lysozyme inhibiting the melanization. They showed that direct protein interaction between lysozyme and pro-PO inhibited its cleavage and therefore the activation pathway; however, lysozyme had no effect on active PO. *Plasmodium* apparently evolved to avoid attacks from *Anopheles* immune system taking advantage of lysozyme interaction.

Kajla et al. (2011) considered the possibility that the regulation of parasite development might offer new target for malaria control. Although this research field may open the possibility to develop malaria control tools, there is not a neat picture of *Plasmodium–Anopheles* interactions yet. The role of lysozymes in the regulation of oocysts development and the mechanism of action are still unclear.

References

Abraham EG, Nagaraju J, Salunke D et al (1995) Purification and partial characterization of an induced antibacterial protein in the silkworm, Bombyx mori. J Invertebr Pathol 65:17–24

Adini A, Warburg A (1999) Interaction of *Plasmodium gallinaceum* ookinetes and oocysts with extracellular matrix proteins. Parasitology 119:331–336

Ahmed AM, Maingon BR, Hurd H (2002) The cost of mounting of an immune response are reflected in the reproductive fitness of the mosquito *Anopheles gambiae*. Oikos 97:371–377

Arrighi RBG, Lycett G, Mahairaki V et al (2005) Laminin and the malaria parasite's journey through the mosquito midgut. J Exp Biol 208:2497–2502

Bachali S, Jager M, Hassanin A et al (2002) Phylogenetic analysis of invertebrate lysozymes and the evolution of lysozyme function. J Mol Evol 54:652–664

Callewaert L, Michiels CW (2010) Lysozymes in the animal kingdom. J Biosci 35:127–160

Castillo JC, Robertson AE, Strand MR (2006) Characterization of hemocytes from the mosquito *Anopheles gambiae* and *Aedes aegypti*. Insect Biochem Mol Biol 36:891–903

Christensen BM, Li J, Chen CC, Nappi AJ (2005) Melanization immune responses in mosquito vectors. Trends Parasitol 21:192–199

Cotter SC, Myatt JP, Benskin CM, Wilson K (2008) Selection for cuticular melanism reveals immune function and life-history trade-offs in *Spodoptera littoralis*. J Evol Biol 21:1744–1754

Daffre S, Kylsten P, Samakovlis C, Hultmark D (1994) The lysozyme locus in *Drosophila melanogaster*: an expanded gene family adapted for expression in the digestive tract. Mol Gen Genet 242:152–162

Dessens JT, Siden-Kiamos I, Mendoza J et al (2003) SOAP, a novel malaria ookinetes protein involved in mosquito midgut invasion and oocyst development. Mol Microbiol 49:319–329

Dong Y, Aguilar R, Xi Z et al (2006) *Anopheles gambiae* immune responses to human and rodent *Plasmodium* parasite species. PLoS Pathog 2:e52

Dong Y, Manfredini F, Dimopoulos G (2009) Implication of the mosquito midgut microbiota in the defense against malaria parasites. PLoS Pathog 5:e1000423

During K, Porsh P, Mahn A (1996) The non-enzymatic microbicidal activity of lysozymes. FEBS Lett 449:93–100

Hultmark D (1996) Insect lysozymes. EXS 75:87–102

Ibrahim HR, Thomas U, Pellegrini A (2001) A helix–loop–helix peptide at the upper lip of the active site cleft of lysozyme confers potent antimicrobial activity with membrane permeabilization action. J Biol Chem 276:43767–43774

Ito Y, Yoshikawa A, Hotani T et al (1999) Amino acid sequences of lysozymes newly purified from invertebrates imply wide distribution of a novel class in the lysozyme family. Eur J Biochem 259:456–461

Kajla MK, Andreeva O, Gilbreath TM et al (2010) Characterization of expression, activity and role in antibacterial immunity of *Anopheles gambiae* lysozyme c-1. Comp Biochem Physiol B Biochem Mol Biol 155:201–209

Kajla MK, Shi L, Li B et al (2011) A new role for an old antimicrobial: lysozyme c-1 can function to protect malaria parasites in *Anopheles* mosquitoes. PLoS One 6:e19649

Kang D, Romans P, Lee JY (1996) Analysis of a lysozyme gene from the malaria vector mosquito, *Anopheles gambiae*. Gene 174:239–244

Kim CH, Park J-W, Ha N-C et al (2008) Innate immune response in insects: recognition of bacterial peptidoglycan and amplification of its recognition signal. BMB Rep 41:93–101

Kylsten P, Kimbrell DA, Daffre S, Samakovlis C, Hultmark D (1992) The lysozyme locus in *Drosophila melanogaster*: different genes are expressed in gut and salivary glands. Mol Gen Genet 232:335–343

Lemos FJA, Ribeiro AF, Terra WR (1993) A bacteria-digesting midgut-lysozyme from *Musca domestica* (diptera) larvae. Purification, properties and secretory mechanism. Insect Biochem Mol Biol 23:533–541

Li B, Calvo E, Marinotti O et al (2005) Characterization of the c-type lysozyme gene family in *Anopheles gambiae*. Gene 360:131–139

Li B, Paskewitz SM (2006) A role for lysozyme in melanization of Sephadex beads in *Anopheles gambiae*. J Insect Physiol 52:936–942

Mahairaki V, Voyatzi T, Siden-Kiamos I, Louis C (2005) The *Anopheles gambiae* gamma 1 laminin directly binds the *Plasmodium berghei* circumsporozoite and TRAP-related protein (CTRP). Mol Biochem Parasitol 140:119–121

Mai W, Hu C (2009) cDNA cloning, expression and antibacterial activity of lysozyme C in the blue shrimp (*Litopanaeus stylirostris*). Prog Nat Sci 19:837–844

Masschalck B, Michiels CW (2003) Antimicrobial properties of lysozyme in relation to foodborne vegetative bacteria. Crit Rev Microbiol 29:191–214

Moss JM, Van Damme MPI, Murphy WH, Preston BN (1997) Dependence of salt concentration on glycosaminoglycan-lysozyme interactions in cartilage. Arch Biochem Biophys 348:49–55

Meis JF, Pool G, van Gemert GJ et al (1989) *Plasmodium falciparum* ookinetes migrate intercellularly through *Anopheles stephensi* midgut epithelium. Parasitol Res 76:13–19

Moreira-Ferro CK, Marinotti O, Bijovsky AT (1999) Morphological and biochemical analyses of the salivary glands of the malaria vector, *Anopheles darlingi*. Tissue Cell 31:264–273

Nacer A, Walker K, Hurd H (2008) Localisation of laminin within *Plasmodium berghei* oocysts and the midgut epithelial cells of *Anopheles stephensi*. Parasit Vectors 1:33

Nakimbugave D, Masschalck B, Atanassova M et al (2006) Comparison of bactericidal activity of six lysozymes at atmospheric pressure and under high hydrostatic pressure. Int J Food Microbiol 108:355–363

Nilsen IW, Overbo K, Sandsdalen E et al (1999) Protein purification and gene isolation of chlamysin, a cold-active lysozyme-like enzyme with antibacterial activity. FEBS Lett 464: 153–158

Osta M, Christophides GK, Kafatos FC (2004) Effects of mosquito genes on *Plasmodium* development. Science 303:2030–2032

Park PW, Biedermann K, Mecham L et al (1996) Lysozyme binds to elastin and protects elastin from elastase-mediated degradation. J Invest Dermatol 106:1075–1080

Park JW, Kim CH, Kim JH et al (2007) Clustering of peptidoglycan recognition protein-SA is required for sensing lysine-type peptidoglycan in insects. Proc Natl Acad Sci U S A 104: 6602–6607

Paskewitz SM, Li B, Kajla MK (2008) Cloning and molecular characterization of two invertebrate-type lysozymes from *Anopheles gambiae*. Insect Mol Biol 17:217–225

Povey S, Cotter SC, Simpson SJ et al (2009) Can the protein costs of bacterial resistance be offset by altered feeding behaviour? J Anim Ecol 78:437–446

Rao X-J, Ling E, Yu X-Q (2010) The role of lysozyme in the prophenoloxidase activation system of *Manduca sexta*: an in vitro approach. Dev Comp Immunol 34:264–271

Reddy JT, Locke M (1990) The size limited penetration of gold particles through insect basal laminae. J Insect Physiol 36:397–408

Rossignol PA, Lueders AM (1986) Bacteriolytic factor in the salivary glands of the *Aedes aegypti*. Comp Biochem Physiol 83:819–822

Skerrett SJ (2004) Lysozyme in pulmonary host defense. Am J Respir Crit Care Med 169: 435–436

Vlachou D, Lycett G, Siden-Kiamos I et al (2001) *Anopheles gambiae* laminin interacts with the P25 surface protein of *Plasmodium berghei* ookinetes. Mol Biochem Parasitol 112:229–237

Yu KH, Kim KN, Lee JH et al (2002) Comparative study on characteristics of lysozymes from the hemolymph of three lepidopteran larvae, *Galleria mellonella*, *Bombyx mori*, *Agrius convolvuli*. Dev Comp Immunol 26:707–713

Zavalova LL, Baskova IP, Lukyanov SA et al (2000) Destabilase from the medicinal leech is a representative of a novel family of lysozymes. Biochim Biophys Acta 1478:69–77

Chapter 5
Effects of Malaria Products on Human Monocyte and Neutrophil Degranulation and Lysozyme Release

Manuela Polimeni, Giuliana Giribaldi, and Mauro Prato

1 Introduction

During the intraerythrocytic developmental stages of *Plasmodium* parasites, the high amounts of hemoglobin are catabolized while toxic heme (ferroprotoporphyrin IX) accumulates in parasite digestive vacuole, where it is detoxified through biocrystallization into malarial pigment hemozoin (Hz) (Egan 2008). Free Hz released into the bloodstream as well as infected red blood cells (RBCs) is avidly phagocytosed by monocytes, macrophages, and neutrophils. As a consequence of phagocytosis, the phenotype of these cells is dramatically altered and their functions are seriously impaired, as schematized in Fig. 5.1 (Schwarzer et al. 1992; Prato and Giribaldi 2011). Interestingly, Hz promotes monocyte degranulation and lysozyme release after inducing the expression and secretion of several proinflammatory molecules (Prato et al. 2009; Polimeni et al. 2012, 2013b). Hz also enhances the release of human matrix metalloproteinase-9 (MMP-9) and tissue inhibitor of metalloproteinase-1 (TIMP-1), two molecules stored in specific gelatinase granules along with lysozyme (Prato et al. 2005, 2008, 2010a; Giribaldi et al. 2010, 2011; Polimeni et al. 2013a; Polimeni and Prato 2014). The present chapter will describe Hz-dependent lysozyme release from human phagocytes, focusing on the underlying mechanisms, as well as on the involved Hz biochemical components.

M. Polimeni (✉) • G. Giribaldi
Dipartimento di Oncologia, Università degli Studi di Torino, Torino, Italy
e-mail: manuela.polimeni@unito.it

M. Prato
Dipartimento di Neuroscienze, Università degli Studi di Torino, Torino, Italy

Dipartimento di Scienze della Sanità Pubblica e Pediatriche, Università degli Studi di Torino, Torino, Italy

M. Prato (ed.), *Human and Mosquito Lysozymes*,
DOI 10.1007/978-3-319-09432-8_5

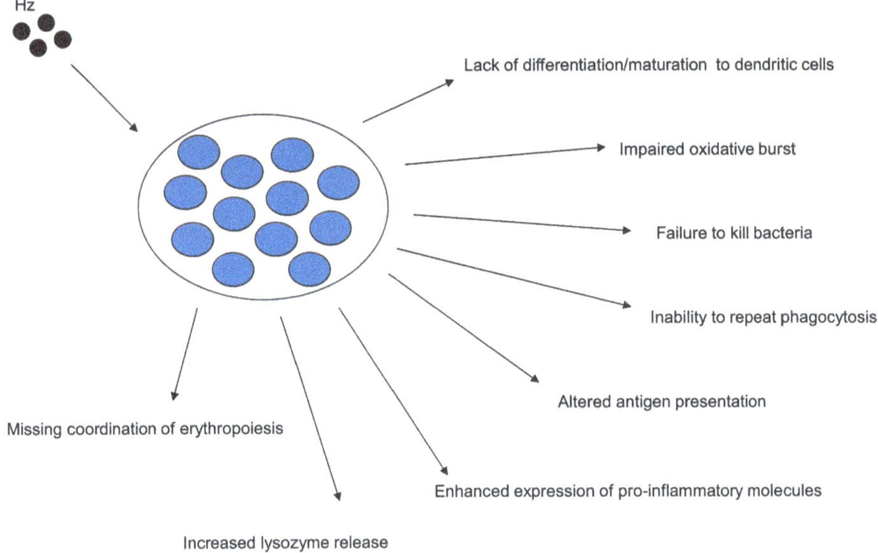

Fig. 5.1 Hz functionally impairs human monocytes and alters their phenotype. The figure shows Hz-dependent impairment of main monocyte functions (oxidative burst, repeated phagocytosis, bacteria killing, antigen presentation, differentiation and maturation to dendritic cells, erythropoiesis) as well as representative phenotype modifications (secretion of proinflammatory molecules)

2 Hemozoin Induces Human Phagocyte Degranulation and Lysozyme Secretion

The malarial pigment Hz, released into blood circulation after infected RBCs rupture at the schizont stage, is rapidly removed by circulating phagocytes such as monocytes and granulocytes. However, upon phagocytosis Hz cannot be degraded and remains in the host for a long time after the infection has been cleared (Shio et al. 2010), thus impairing the functionality of human phagocytes (see Fig. 5.1). For example, Hz-fed monocytes are unable to mount oxidative burst and repeat phagocytosis (Schwarzer et al. 1992), do not kill ingested bacteria (Fiori et al. 1993), fail presenting antigens correctly (Scorza et al. 1999), do not differentiate and mature to dendritic cells (Urban and Todryk 2006), and do not properly coordinate erythropoietic events (Giribaldi et al. 2004). Moreover, the phenotype of human monocytes is dramatically altered, as these cells shed large amounts of bioactive peroxidation derivatives of polyunsaturated fatty acids (Schwarzer et al. 2003) and produce enhanced amounts of a large number of proinflammatory and/or anti-apoptotic molecules, including cytokines (IL-1β, TNF-α, IL-1RA), chemokines (MIP-1α/CCL3, MIP-1β/CCL4, MCP-1/CCL2, IL-8/CXCL8, ENA-78/CXCL5, GROα/CXCL1, GROβ/CXCL2, GROγ/CXCL3) (Pichyangkul et al. 1994; Sherry et al. 1995; Giribaldi et al. 2010), heat shock proteins (HSP-27) (Prato et al. 2010b),

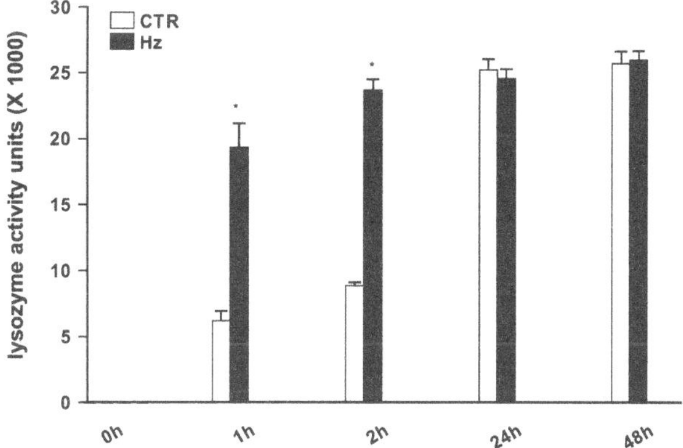

Fig. 5.2 Early time-dependent induction of lysozyme release from Hz-fed human adherent monocytes. Cells were left unfed (control cells, CTR) or fed with natural Hz for 2 h. Thereafter, lysozyme levels in cell supernatants were monitored 0, 1, 2, 24, and 48 h after the end of phagocytosis by spectrometric assay. Data are means+SEM of three independent experiments. Lysozyme release from monocytes is indicated as enzyme activity units measured in a 2-mL well of cell supernatants. All data were evaluated for significance by ANOVA: versus CTR $*p<0.0001$ (Adapted from Polimeni et al. 2012)

lytic enzymes (MMP-9, lysozyme) (Prato et al. 2005, 2009), and enzyme inhibitors (TIMP-1) (Polimeni et al. 2013a) (see Fig. 5.1).

Hz-dependent monocyte degranulation is associated with enhanced release of lysozyme, stored in the so-called gelatinase granules. Indeed, it has been recently demonstrated in vitro that phagocytosis of Hz—either free or contained in trophozoites—by human adherent monocytes induces an early increase in the release of active lysozyme (Prato et al. 2009). Hz-increased lysozyme levels appear significantly higher than those induced by phagocytosis of opsonized RBCs or latex particles, thus suggesting that Hz effects are not dependent on phagocytosis per se. Interestingly, as shown in Fig. 5.2, Hz promotes lysozyme release quite immediately (1–2 h after phagocytosis), whereas at later time points (24–48 h) a plateau is reached, possibly as a result of complete degranulation of human monocytes (Polimeni et al. 2012). Intriguingly, since Hz-induced degranulating response is very rapid and unlikely accompanied by any second waves, lysozyme could be a potential suitable marker for early diagnosis of severe malaria (Prato 2012).

3 Hz-Dependent Lysozyme Release Is Mediated by TNF-α, IL-1β, and MIP-1α/CCL3

The possible soluble mediators responsible for Hz-dependent lysozyme release (see Fig. 5.3) have been investigated in a series of recent studies (Prato et al. 2009; Polimeni et al. 2012, 2013b). TNF-α was the first mediator to be evaluated, showing

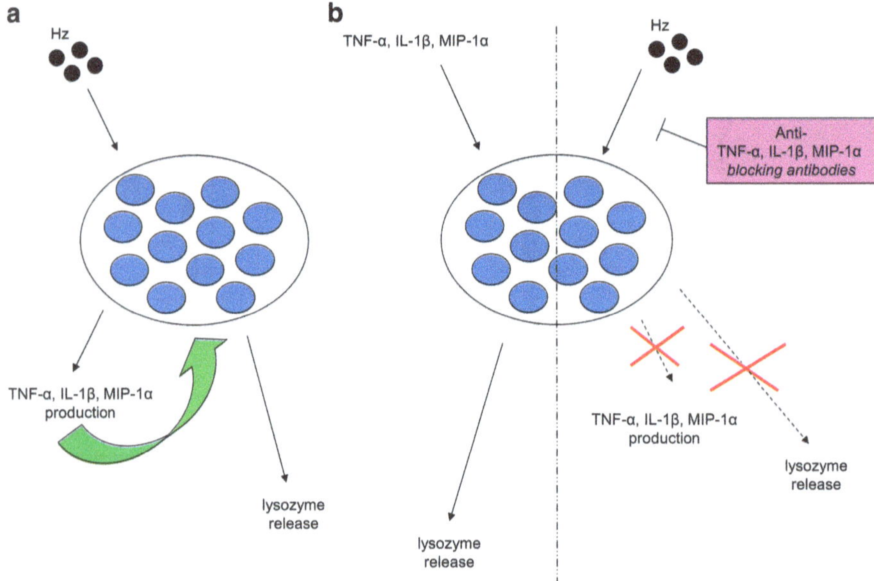

Fig. 5.3 Involvement of proinflammatory molecules (TNF-α, IL-1β, MIP-1α/CCL3) in Hz-dependent lysozyme release. (**a**) Shows the effects of natural Hz on production and secretion of several proinflammatory molecules (TNF-α, IL-1β, MIP-1α/CCL3) by human monocytes. In (**b**) the experiments with recombinant cytokines (on the *left*) and blocking antibodies (on the *right*) performed to demonstrate TNF-α, IL-1β, MIP-1α/CCL3 involvement in upregulation of lysozyme release are schematized

an increase up to 72 h after Hz phagocytosis (Prato et al. 2005). Inhibition with anti-TNF-α blocking antibodies or stimulation with the recombinant exogenous cytokine revealed that higher levels of soluble TNF-α were actively involved in Hz-triggered increase of lysozyme release from human adherent monocytes. Indeed, the upregulating effects of Hz were mimicked by exogenous recombinant human TNF-α and abrogated by blocking anti-TNF-α antibodies (Prato et al. 2009). In human phagocytes such as neutrophil granulocytes and monocytes, lysozyme is stored in different types of granules, including the so-called gelatinase granules, which also contain MMP-9 (Borregaard 1997). Consistently, Hz was previously shown to enhance the expression, release and activity of human constitutive monocytic MMP-9 (Prato et al. 2005) but not inducible MMP-2 (Prato 2011). Such an enhancement appears to be mediated by several proinflammatory molecules, including TNF-α (Prato et al. 2005), IL-1β (Prato et al. 2008), and MIP-1α/CCL3 (Giribaldi et al. 2011). The gene expression of these molecules, along with that of other cytokines and chemokines, is also upregulated by Hz (Pichyangkul et al. 1994; Sherry et al. 1995; Giribaldi et al. 2010). Praecox IL-1β production has been suggested to trigger the expression of all other genes, as suggested by results from a macroarray study (Giribaldi et al. 2010). This evidence is strengthened by some data emerging from further investigation on protein secretion, showing that Hz-enhanced protein release of IL-1β and MIP-1α/CCL3, was already significant 1 h after the end of phagocytosis, whereas TNF-α was significantly increased 120 min after phagocytosis (Polimeni et al. 2012, 2013b). As resulting from experiments with blocking antibodies and recombinant cytokines, IL-1β and MIP-1α/CCL3 appeared to be causally connected to Hz-dependent upregulation of lysozyme release in a manner similar to TNF-α. Indeed, anti-IL-1β and anti-MIP-1α/CCL3 blocking antibodies reduced—although did not fully abrogate—Hz-dependent induction of lysozyme release, while recombinant IL-1β and MIP-1α/CCL3 partially mimicked the effects of Hz, promoting lysozyme release without reaching the levels induced by Hz. Apparently, all three molecules are required as soluble mediators to fulfill Hz-dependent lysozyme upregulation, since the combination of all three exogenous proinflammatory molecules was able to trigger lysozyme levels similar to those induced by Hz alone while a mix of all the blocking antibodies fully abrogated the effect of Hz. These data are consistent with previous documents showing lysozyme gene dependence on cytokine levels (Berger et al. 1988; Lewis et al. 1990).

Nevertheless, it should be underlined that, contrary to lysozyme, Hz-dependent enhanced production of TNF-α, IL-1β, and MIP-1α/CCL3 appears continuous, going on also at longer times: for instance, in Hz-fed monocyte supernatants 25 ng/mL TNF-α was measured 72 h after phagocytosis (Prato et al. 2005); 35 ng/mL IL-1β was found 48 h after phagocytosis (Prato et al. 2008); and 20 ng/mL MIP-1α/CCL3 was retrieved 24 h after phagocytosis (Giribaldi et al. 2011). Moreover, it has been reported that Hz-fed monocytes produce increased amounts of several chemotactic molecules other than MIP-1α/CCL3 (including MIP-1β/CCL4, GROα/CXCL1, GROβ/CXCL2, GROγ/CXCL3, MCP-1/CCL2, IL-8/CXCL8, and ENA-78/CXCL5) (Giribaldi et al. 2010) and show compromised ability to perform repeated phagocytosis (Schwarzer et al. 1992). Thus, in order to translate the present findings into acute and chronic human malaria and to explain the higher plasma levels of lysozyme found in patients with malaria (Mohamed et al. 1996; Mohammed et al. 2003), it is reasonable to hypothesize that the lysozyme accumulation might be a consequence of two complementary events, as schematized in Fig. 5.4: on the one hand, immediately after Hz phagocytosis human monocytes may release higher amounts of lysozyme; on the other hand, after full release of lysozyme, Hz-laden monocytes could go on producing proinflammatory molecules, such as TNF-α, IL-1β, and MIP-1α/CCL3, as well as other chemotactic molecules, which might in turn recruit more monocytes into the areas of parasite sequester in microvessels. Therefore, as a consequence of increased circulating levels of TNF-α, IL-1β, and MIP-1α/CCL3 and as a result of additional phagocytosis of Hz by newly recruited monocytes (but not of repeated phagocytosis by those already laden), new lysozyme release would be induced, thus contributing to enhance total lysozyme circulating levels (Polimeni et al. 2012) (Fig. 5.3).

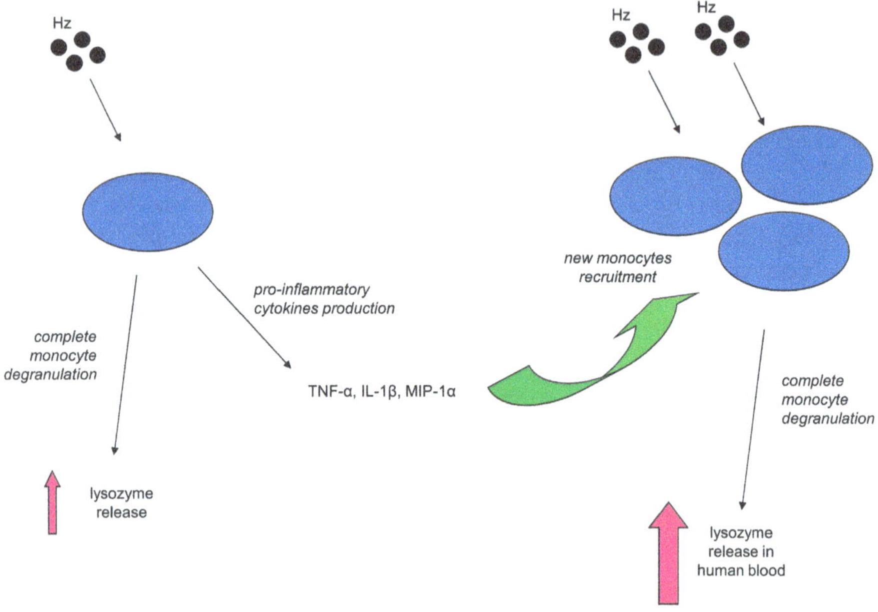

Fig. 5.4 Putative mechanism underlying lysozyme release and accumulation in the blood of human malaria patients. The scheme illustrates the two complementary events hypothesized to account for new lysozyme production and release and its following accumulation in the blood of human malaria patients: human monocytes can release higher amounts of lysozyme after Hz phagocytosis (on the *left*); after full release of lysozyme, Hz-laden monocytes can go on producing proinflammatory molecules able to recruit more monocytes into the areas of parasite sequester in the microvessels, thereby contributing to enhance total lysozyme circulating levels

4 Hz Induces Activation of p38 MAPK and NF-κB Pathways

Once demonstrated Hz-enhanced lysozyme release from human granulocytes, the putative mechanisms of signal transduction underlying such an enhancement have been investigated, mainly focusing on p38 mitogen-activated protein kinase (MAPK) and nuclear factor-kappaB (NF-κB) pathways (Polimeni et al. 2012).

The pathways activated by Hz in human monocytes have been scarcely described in the past years. However, some helpful information comes from a few in vitro and in vivo studies, suggesting an involvement of MAPKs in malaria. In murine macrophages or monocytes, Hz was shown to induce activation of p38 (Cambos et al. 2010) and ERK1/2 (Jaramillo et al. 2003, 2005; Griffith et al. 2009) but not JNK-2/STAT-1 (Jaramillo et al. 2003, 2005) MAPK pathways, whereas *P. falciparum* glycosylphosphatidylinositol (*Pf*GPI) promoted phosphorylation of all routes (Lu et al. 2006; Zhu et al. 2009). Interestingly, inhibition of *Pf*GPI-dependent activation of MAPK decreased inflammatory responses and enhanced phagocytic clearance of infected RBCs in mice infected by *P. berghei* or *chabaudi* (Serghides et al. 2009). Lucchi and colleagues reported in human syncytiotrophoblast cells a Hz-dependent phosphorylation of ERK1/2 and an infected RBC-dependent phosphorylation of JNK-1; both events were causally related to the production of pro-inflammatory molecules (TNF-α, MIP-1α/CCL3, IL-8/CXCL8) (Lucchi et al. 2008, 2011).

In human monocytes, it has been recently shown that Hz activates p38 MAPK pathway by inducing p38 MAPK phosphorylation without altering basal protein levels (Polimeni et al. 2012). Such an event is directly connected to Hz-induced lysozyme levels, as demonstrated by experiments using the specific synthetic inhibitor of p38 signaling SB203580, which abrogated Hz enhancing effects on lysozyme release into monocyte supernatants in vitro. These observations are consistent with previous data correlating activation of MAPK pathways to degranulation from human neutrophils (Sue-A-Quan et al. 1997) and p38 MAPK phosphorylation to lysozyme release in vibrio-infected mussel granulocytes (Ciacci et al. 2010).

In addition, *c*-lysozyme gene has been reported to be regulated also through NF-κB signaling (van Phi 1996). Interestingly, evidence on Hz-dependent NF-κB activation is presently available: in human monocytes, natural Hz was shown to upregulate MMP-9 through long-term (24 h after phagocytosis) I-κBα degradation and NF-κB nuclear translocation in vitro (Dell'agli et al. 2010; Prato et al. 2010a), whereas in murine macrophages either natural or synthetic Hz promoted cytokine and chemokine production through NF-κB activation both in vitro and in vivo (Jaramillo et al. 2003, 2005; Griffith et al. 2009). Proof of early activation of NF-κB pathway has also been recently documented, as Hz was shown to induce cytosolic I-κBα phosphorylation and degradation, p50/p65 NF-κB subunits nuclear translocation, and NF-κB/DNA binding 2 h after phagocytosis by human monocytes (Polimeni et al. 2012). All these effects of Hz were abrogated by quercetin, artemisinin, and parthenolide, three molecules reported as NF-κB inhibitors (García-Piñeres et al. 2001; Aldieri et al. 2003; Nair et al. 2006) and showing antimalarial properties (Khalid et al. 1986; Burrows et al. 2011). Moreover, quercetin, artemisinin, and parthenolide inhibited Hz-dependent enhancement of lysozyme levels in monocyte supernatants, suggesting a potential role for NF-κB in lysozyme regulation by Hz. Interestingly, in a previous work these three inhibitors abrogated the upregulating effects of Hz on MMP-9, TNF-α, and IL-1β production by human monocytes (Prato et al. 2010a).

Collectively, natural Hz-enhanced lysozyme release from human monocytes appears to be dependent on overproduction of TNF-α, IL-1β, and MIP-1α/CCL3 and on activation of p38 MAPK and NF-κB pathways. For more clearness, the overall mechanism underlying Hz-triggered increase of lysozyme release has been schematized in Fig. 5.5.

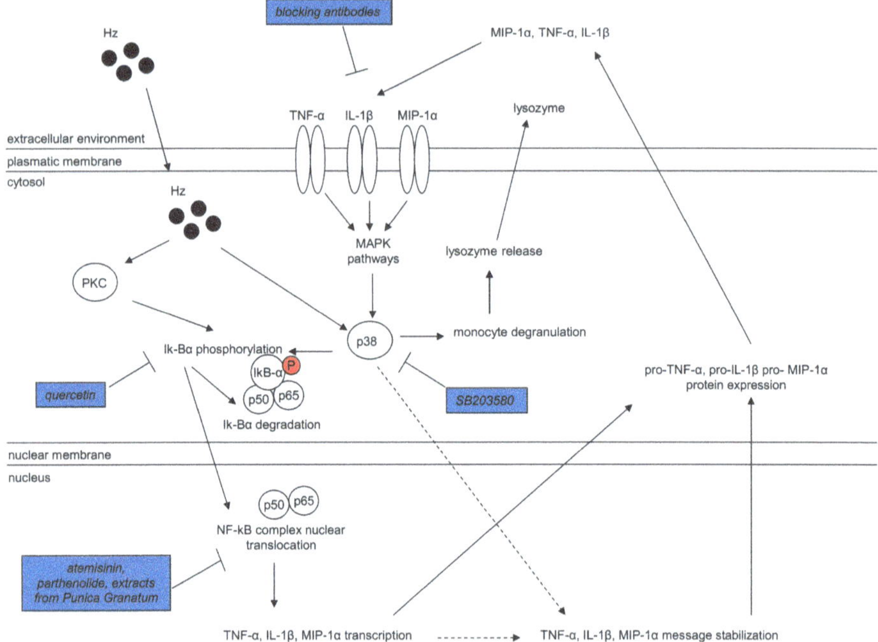

Fig. 5.5 Signaling pathways activated by Hz in human monocytes. The figure schematizes the possible mechanisms underlying Hz-enhanced lysozyme release in human monocytes. Immediately after phagocytosis, Hz activates PKC and p38 MAPK. These kinases have been associated with cytosolic I-κBα phosphorylation and degradation, resulting in subsequent nuclear translocation of NF-κB p50 and p65 subunits. As a consequence of NF-κB pathway activation, the transcription and protein expression of several proinflammatory molecules including TNF-α, IL-1β, and MIP-1α/CCL3 is promoted. Thereafter, these molecules can induce in turn lysozyme release into the extracellular environment via p38 MAPK

5 15-HETE Plays a Major Role in Hz-Dependent Enhancing Effects on Lysozyme Release

Natural Hz has a scaffold structure composed both by the ferric heme and by the lipid moiety, which contains large amounts of lipoperoxidation products generated by nonenzymatic catalysis, including monohydroxy-derivatives of arachidonic and linoleic acids [13- and 9-hydoxy-octadecadienoic (HODEs) acids; 9-, 12-, and 15-hydroxyeicosatetraenoic (HETE) acids] and 4-hydroxynonenal (4-HNE), stable end products of hydroxylation and peroxidation of polyenoic fatty acids (Schwarzer et al. 2003). 15-HETE and 4-HNE are directly related to functional impairment of nHz-fed human monocytes, whereas HODEs are almost inert (Prato and Giribaldi 2011) (see Fig. 5.6a).

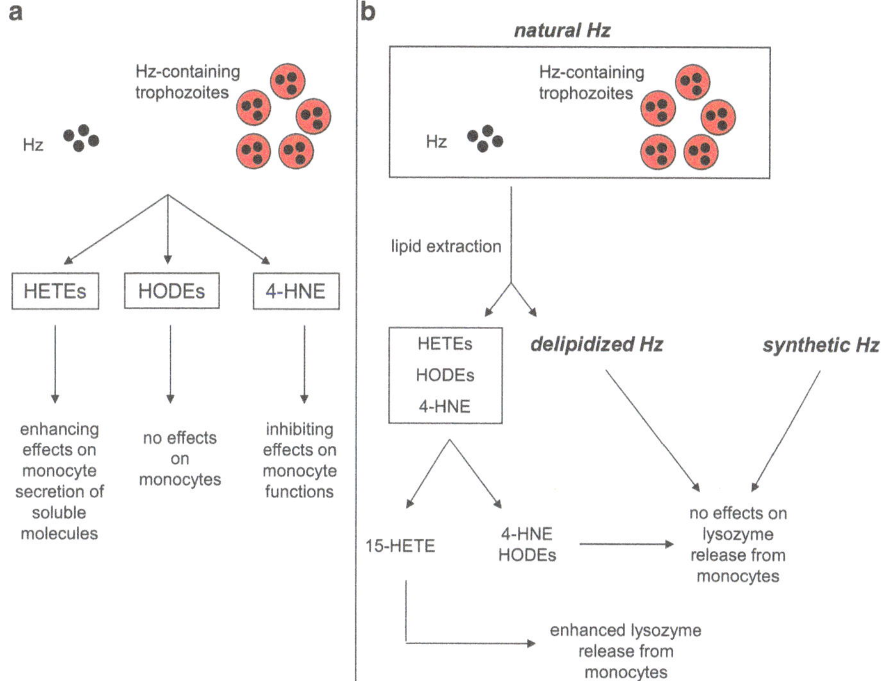

Fig. 5.6 15-HETE reproduces natural Hz-dependent effects on lysozyme release. (**a**) Schematizes the different components (HETEs, HODEs, and 4-HNE) of the lipid moiety of natural Hz (either free or contained in trophozoites), highlighting their effects on human monocytes. (**b**) Shows the different effects of natural, delipidized, synthetic Hz along with the different components of Hz lipid moiety on lysozyme release from human monocytes

Lipid-free synthetic Hz and delipidized Hz do not reproduce the effects of natural Hz on lysozyme early release, suggesting a major role for the lipid moiety of Hz (Polimeni et al. 2012) (see Fig. 5.6b). Nevertheless, it has been recently proposed that the phagocytosis of the packaging digestive vacuole—and not only of Hz released after schizogony—might be at the root of pathway activation in phagocytic cells (Dasari et al. 2011). Thus, an in vivo involvement of the full digestive vacuole as a causative agent of the enhanced lysozyme levels found in plasma of patients with malaria (Mohamed et al. 1996; Mohammed et al. 2003) cannot be excluded at the moment.

In a recent work human adherent monocytes were treated with different concentrations (0.1–10 µM) of 15-HETE and 4-HNE, two major components of the lipid moiety which have been previously associated with nHz-dependent phenotype alteration and functional impairment of these cells (Polimeni et al. 2013b). There are extensive data from literature providing strong evidence that several HETEs, including 5-, 12- and 15-HETE, promote degranulation and lysozyme release from polymorphonuclear leukocytes (Stenson and Parker 1980; Goetzl and Pickett 1980; Bokoch and Reed 1981; Smith et al. 1981; O'Flaherty and Thomas 1985).

In the above-mentioned study, 15-HETE was the only molecule able to reproduce natural Hz effects on lysozyme release. Indeed, at lower concentrations (0.1–1 μM) it promoted dose dependently lysozyme secretion from human monocytes, whereas a plateau was apparently reached after using the highest one (10 μM); on the contrary, any concentrations of 4-HNE did not mimic nHz effects on lysozyme release (see Fig. 5.6b). This is in general agreement with the hypothesis that 15-HETE plays a major role in natural Hz-dependent enhancing effects on expression and secretion of proinflammatory molecules by human monocytes, whereas 4-HNE is more likely to be involved in impairing the immune functions of these cells (Prato and Giribaldi 2011) (see Fig. 5.6a).

As previously mentioned, phagocytosis of natural Hz by human monocytes promotes IL-1β, TNF-α, and MIP-1α/CCL3 expression and secretion (Prato et al. 2005, 2008, 2010a; Giribaldi et al. 2010, 2011), and early secretion of these molecules mediates Hz-dependent induction of lysozyme release (Polimeni et al. 2012). IL-1β, TNF-α, and MIP-1α/CCL3 have also been reported as soluble mediators of nHz-enhanced expression and release of MMP-9 (Prato et al. 2005, 2008; Giribaldi et al. 2011) and TIMP-1 (Polimeni et al. 2013a). The effects of the differential components of nHz on lysozyme release, as well as the possible soluble mediators and underlying mechanisms, have been explored in a recent study (Polimeni et al. 2013b).

Apparently, the lipid moiety of natural Hz is required to early enhance the production of IL-1β and TNF-α, but not that of MIP-1α/CCL3, whose upregulation seems dependent on heme crystal through unknown mechanisms. Interestingly, 15-HETE was shown to be involved in Hz-dependent early production of IL-1β and TNF-α, whereas 4-HNE was not. These data are consistent with previous results from either short-term studies, showing that 15-HETE does not alter MIP-1α/CCL3 mRNA expression 6 h after treatment of human immunopurified monocytes (Giribaldi et al. 2010), or long-term studies, describing higher protein levels of IL-1β and TNF-α (along with those of MMP-9) 48 h after treatment with 15-HETE of human adherent monocytes (Prato et al. 2008, 2010a). Interestingly, 12-HETE, a peroxidated lipid closely related to 15-HETE, was reported to stimulate TNF-α mRNA and protein expression in a dose-dependent manner in murine macrophages (Wen et al. 2007) and to enhance IL-1β-stimulated nitrite production and iNOS expression in vascular smooth muscle cells (Hashimoto et al. 2003), thereby confirming that HETEs can upregulate the production of both cytokines. On the other hand, MIP-1α expression was upregulated in murine macrophages fed with either natural Hz or lipid-free synthetic Hz, confirming that lipids are irrelevant in nHz-dependent enhancement of MIP-1α/CCL3 (Jaramillo et al. 2005). Since these three molecules were suggested to be involved in natural Hz-dependent enhancement of lysozyme (Polimeni et al. 2012), a double blocking/mimicking approach with neutralizing antibodies and recombinant cytokines/chemokines was used in order to confirm their possible role as soluble mediators of lysozyme regulation by 15-HETE. IL-1β and TNF-α appeared to be causally connected to the 15-HETE-dependent induction of lysozyme release, whereas MIP-1α/CCL3 did not. Therefore, since MIP-1α/CCL3 is upregulated by natural Hz regardless of its lipid moiety and does not concur in 15-HETE-dependent lysozyme release induction, it appears reasonable to speculate a redundant role for this chemokine in achieving natural Hz effects on lysozyme secretion (Fig. 5.5). Besides, previous documents reported lysozyme gene dependence on levels of cytokines (Berger et al. 1988; Lewis et al. 1990).

The mechanisms of signal transduction promoted by 15-HETE, along with its involvement in natural Hz-induced lysozyme release, have also been investigated (Polimeni et al. 2013b). Activation of p38 MAPK and NF-κB pathways were previously suggested to be two integrated mechanisms underlying natural Hz-dependent upregulation of lysozyme levels secreted by human monocytes (Polimeni et al. 2012). Interestingly, activation of p38 MAPK pathway by HETEs has been reported in several cells: 15-HETE induces p38 MAPK phosphorylation in rat smooth muscle cells (Chava et al. 2009) and human retinal microvascular endothelial cells (Bajpai et al. 2007); additionally, 12-HETE-dependent phosphorylation has been observed in murine macrophages (Wen et al. 2007, 2008), murine fibroblasts (Nieves and Moreno 2008), and human adrenocortical cells (Natarajan et al. 2002). Consistently, in human monocytes 15-HETE was shown to promote early phosphorylation of p38 MAPK without affecting basal protein levels, similarly to natural Hz. Since chemical inhibition of this event abrogated 15-HETE-induced lysozyme release, 15-HETE was proposed to play a major role in natural Hz enhancing effects on monocyte degranulation (Polimeni et al. 2013b).

Natural Hz also promotes both early (Prato et al. 2010a) and late (Polimeni et al. 2012) activation of NF-κB pathway in human monocytes by inducing phosphorylation/degradation of cytosolic I-κBα and subsequent NF-κB nuclear translocation/DNA binding. Long-term activation of NF-κB transcription system by natural Hz was mimicked by 15-HETE, whereas chemical inhibition of this route reduced the long-term increased levels of MMP-9, IL-1β, and TNF-α (Prato et al. 2010a). 15-HETE has also been proposed to be responsible for early activation of NF-κB pathway in human monocytes, and such an event appears to be directly related to enhanced lysozyme levels, as demonstrated by blocking experiments with three different NF-κB inhibitors (Polimeni et al. 2013b). According to literature, both inhibition and activation of NF-κB by HETEs have been described. HETEs and parent compounds HODEs can bind peroxisome proliferator-activated receptor-γ (PPAR-γ) (Huang et al. 1999). Thus, they should be expected to suppress the NF-κB system (Cabrero et al. 2002). However, HETEs can also activate NF-κB through several alternative mechanisms, thereby contrasting PPAR-γ effects. Above all, it has been convincingly suggested that HETE-dependent activation of MAPK pathways, such as ERK and p38 MAPK, leads to the concurrent activation of NF-κB system (Di Mari et al. 2007; Ishizuka et al. 2008; Chen et al. 2008; Cheng et al. 2010). Thus, 15-HETE-dependent NF-κB activation in nHz-fed human monocytes appears reasonably justified by concurrent 15-HETE-dependent p38 phosphorylation, as previously described. Additionally, HETEs can activate protein kinase C (Sharma et al. 2005; Wyke et al. 2005; Chen et al. 2008), which in turn activates IKK (Smith et al. 2004), the kinase responsible for I-κBα phosphorylation and subsequent activation of NF-κB pathway (Verma and Stevenson 1997). Interestingly, transient early activation of protein kinase C by natural Hz has been reported (Schwarzer et al. 1993). Finally, activation of NF-κB may result from the direct oxidation and inactivation of I-κBα by lipid hydroperoxides (Flohé et al. 1997).

6 Conclusion

Recent findings showed natural Hz-dependent upregulation of lysozyme release in human monocytes. Such increase was mediated by Hz-induced proinflammatory cytokines, including IL-1β, TNF-α, and MIP-1α/CCL-3 through activation of p38 MAPK and NF-κB transcriptional pathways. In addition, recent evidence strongly supports 15-HETE, the major component of the lipid moiety of natural Hz, as the main actor in these events. Therefore, as discussed in the next chapter, lysozyme could reveal itself as a good marker to design future specific cost-effective diagnostic approaches to detect severe malaria.

References

Aldieri E, Atragene D, Bergandi L et al (2003) Artemisinin inhibits inducible nitric oxide synthase and nuclear factor NF-kB activation. FEBS Lett 552:141–144

Bajpai AK, Blaskova E, Pakala SB et al (2007) 15(S)-HETE production in human retinal microvascular endothelial cells by hypoxia: novel role for MEK1 in 15(S)-HETE induced angiogenesis. Invest Ophthalmol Vis Sci 48(11):4930–4938

Berger M, Wetzler EM, Wallis RS (1988) Tumor necrosis factor is the major monocyte product that increases complement receptor expression on mature human neutrophils. Blood 71: 151–158

Bokoch GM, Reed PW (1981) Effect of various lipoxygenase metabolites of arachidonic acid on degranulation of polymorphonuclear leukocytes. J Biol Chem 256(11):5317–5320

Borregaard N (1997) Development of neutrophil granule diversity. Ann N Y Acad Sci 832:62–68

Burrows JN, Chibale K, Wells TN (2011) The state of the art in anti-malarial drug discovery and development. Curr Top Med Chem 11:1226–1254

Cabrero A, Laguna JC, Vázquez M (2002) Peroxisome proliferator-activated receptors and the control of inflammation. Curr Drug Targets Inflamm Allergy 1(3):243–248

Cambos M, Bazinet S, Abed E et al (2010) The IL-12p70/IL-10 interplay is differentially regulated by free heme and hemozoin in murine bone-marrow-derived macrophages. Int J Parasitol 40: 1003–1012

Chava KR, Karpurapu M, Wang D et al (2009) CREB-mediated IL-6 expression is required for 15(S)-hydroxyeicosatetraenoic acid-induced vascular smooth muscle cell migration. Arterioscler Thromb Vasc Biol 29(6):809–815

Chen F, Wang X, Li J et al (2008) 12-lipoxygenase induces apoptosis of human gastric cancer AGS cells via the ERK1/2 signal pathway. Dig Dis Sci 53:181–187

Cheng J, Wu CC, Gotlinger KH et al (2010) 20-hydroxy-5,8,11,14-eicosatetraenoic acid mediates endothelial dysfunction via IkappaB kinase-dependent endothelial nitric-oxide synthase uncoupling. J Pharmacol Exp Ther 332(1):57–65

Ciacci C, Betti M, Canonico B et al (2010) Specificity of anti-Vibrio immune response through p38 MAPK and PKC activation in the hemocytes of the mussel Mytilus galloprovincialis. J Invertebr Pathol 105:49–55

Dasari P, Reiss K, Lingelbach K et al (2011) Digestive vacuoles of Plasmodium falciparum are selectively phagocytosed by and impair killing function of polymorphonuclear leukocytes. Blood 118:4946–4956

Dell'agli M, Galli GV, Bulgari M et al (2010) Ellagitannins of the fruit rind of pomegranate (Punica granatum) antagonize in vitro the host inflammatory response mechanisms involved in the onset of malaria. Malar J 9:208

Di Mari JF, Saada JI, Mifflin RC et al (2007) HETEs enhance IL-1-mediated COX-2 expression via augmentation of message stability in human colonic myofibroblasts. Am J Physiol Gastrointest Liver Physiol 293:G719–G728

Egan TJ (2008) Haemozoin formation. Mol Biochem Parasitol 157:127–136

Fiori PL, Rappelli P, Mirkarimi SN et al (1993) Reduced microbicidal and anti-tumour activities of human monocytes after ingestion of *Plasmodium falciparum* infected red blood cells. Parasite Immunol 15:647–655

Flohé L, Brigelius-Flohé R, Saliou C et al (1997) Redox regulation of NF-kappa B activation. Free Radic Biol Med 22:1115–1126

García-Piñeres AJ, Castro V, Mora G et al (2001) Cysteine 38 in p65/NF-kappaB plays a crucial role in DNA binding inhibition by sesquiterpene lactones. J Biol Chem 276:39713–39720

Giribaldi G, Ulliers D, Schwarzer E et al (2004) Hemozoin- and 4-hydroxynonenal-mediated inhibition of erythropoiesis. Possible role in malarial dyserythropoiesis and anemia. Haematologica 89:492–493

Giribaldi G, Prato M, Ulliers D et al (2010) Involvement of inflammatory chemokines in survival of human monocytes fed with malarial pigment. Infect Immun 78:4912–4921

Giribaldi G, Valente E, Khadjavi A et al (2011) Macrophage inflammatory protein-1alpha mediates matrix metalloproteinase-9 enhancement in human adherent monocytes fed with malarial pigment. Asian Pac J Trop Med 4:925–930

Goetzl EJ, Pickett WC (1980) The human PMN leukocyte chemotactic activity of complex hydroxy-eicosatetraenoic acids (HETEs). J Immunol 125(4):1789–1791

Griffith JW, Sun T, McIntosh MT et al (2009) Pure Hemozoin is inflammatory in vivo and activates the NALP3 inflammasome via release of uric acid. J Immunol 183:5208–5220

Hashimoto T, Kihara M, Yokoyama K et al (2003) Lipoxygenase products regulate nitric oxide and inducible nitric oxide synthase production in interleukin-1beta stimulated vascular smooth muscle cells. Hypertens Res 26(2):177–184

Huang JT, Welch JS, Ricote M et al (1999) Interleukin-4-dependent production of PPAR-gamma ligands in macrophages by 12/15-lipoxygenase. Nature 400(6742):378–382

Ishizuka T, Cheng J, Singh H et al (2008) 20-Hydroxyeicosatetraenoic acid stimulates nuclear factor-kappaB activation and the production of inflammatory cytokines in human endothelial cells. J Pharmacol Exp Ther 324(1):103–110

Jaramillo M, Gowda DC, Radzioch D et al (2003) Hemozoin increases IFN-gamma-inducible macrophage nitric oxide generation through extracellular signal-regulated kinase- and NF-kappa B-dependent pathways. J Immunol 171:4243–4253

Jaramillo M, Godbout M, Olivier M (2005) Hemozoin induces macrophage chemokine expression through oxidative stress-dependent and -independent mechanisms. J Immunol 174:475–484

Khalid SA, Farouk A, Geary TG et al (1986) Potential antimalarial candidates from African plants: and in vitro approach using Plasmodium falciparum. J Ethnopharmacol 15:201–209

Lewis CE, McCarthy SP, Lorenzen J et al (1990) Differential effects of LPS, IFN-gamma and TNF alpha on the secretion of lysozyme by individual human mononuclear phagocytes: relationship to cell maturity. Immunology 69:402–408

Lu Z, Serghides L, Patel SN et al (2006) Disruption of JNK2 decreases the cytokine response to Plasmodium falciparum glycosylphosphatidylinositol in vitro and confers protection in a cerebral malaria model. J Immunol 177:6344–6352

Lucchi NW, Peterson DS, Moore JM (2008) Immunologic activation of human syncytiotrophoblast by Plasmodium falciparum. Malar J 7:42

Lucchi NW, Sarr D, Owino SO et al (2011) Natural hemozoin stimulates syncytiotrophoblast to secrete chemokines and recruit peripheral blood mononuclear cells. Placenta 32:579–585

Mohamed AO, Elbashir MI, Ibrahim G et al (1996) Neutrophil leucocyte activation in severe malaria. Trans R Soc Trop Med Hyg 90:277

Mohammed AO, Elghazali G, Mohammed HB et al (2003) Human neutrophil lipocalin: a specific marker for neutrophil activation in severe Plasmodium falciparum malaria. Acta Trop 87: 279–285

Nair MP, Mahajan S, Reynolds JL et al (2006) The flavonoid quercetin inhibits proinflammatory cytokine (tumor necrosis factor alpha) gene expression in normal peripheral blood mononuclear cells via modulation of the NF-kappa beta system. Clin Vaccine Immunol 13:319–328

Natarajan R, Yang DC, Lanting L et al (2002) Key role of P38 mitogen-activated protein kinase and the lipoxygenase pathway in angiotensin II actions in H295R adrenocortical cells. Endocrine 18(3):295–301

Nieves D, Moreno JJ (2008) Enantioselective effect of 12(S)-hydroxyeicosatetraenoic acid on 3T6 fibroblast growth through ERK 1/2 and p38 MAPK pathways and cyclin D1 activation. Biochem Pharmacol 76(5):654–661

O'Flaherty JT, Thomas MJ (1985) Effect of 15-lipoxygenase-derived arachidonate metabolites on human neutrophil degranulation. Prostaglandins Leukot Med 17(2):199–212

Pichyangkul S, Saengkrai P, Webster HK (1994) *Plasmodium falciparum* pigment induce monocytes to release high levels of tumor necrosis factor-α and interleukin-1β. Am J Trop Med Hyg 51:430–435

Polimeni M, Prato M (2014) Host matrix metalloproteinases in cerebral malaria: new kids on the block against blood-brain barrier integrity? Fluids Barriers CNS 11(1):1

Polimeni M, Valente E, Aldieri E et al (2012) Haemozoin induces early cytokine-mediated lysozyme release from human monocytes through p38 MAPK- and NF-kappaB-dependent mechanisms. PLoS One 7(6):e39497

Polimeni M, Valente E, Ulliers D et al (2013a) Natural haemozoin induces expression and release of human monocyte tissue inhibitor of metalloproteinase-1. PLoS One 8(8):e71468

Polimeni M, Valente E, Aldieri E et al (2013b) Role of 15-hydroxyeicosatetraenoic acid in hemozoin-induced lysozyme release from human adherent monocytes. Biofactors 39(3): 304–314

Prato M (2011) Malarial pigment does not induce MMP-2 and TIMP-2 protein release by human monocytes. Asian Pac J Trop Med 4:756

Prato M (2012) Human and mosquito lysozymes in malaria: old molecules for new approaches towards diagnosis, therapy and vector control. J Bacteriol Parasitol 3:e113

Prato M, Giribaldi G (2011) Matrix metalloproteinase-9 and haemozoin: wedding rings for human host and Plasmodium falciparum parasite in complicated malaria. J Trop Med 2011:628435

Prato M, Giribaldi G, Polimeni M et al (2005) Phagocytosis of hemozoin enhances matrix metalloproteinase-9 activity and TNF-alpha production in human monocytes: role of matrix metalloproteinases in the pathogenesis of falciparum malaria. J Immunol 175:6436–6442

Prato M, Gallo V, Giribaldi G et al (2008) Phagocytosis of haemozoin (malarial pigment) enhances metalloproteinase-9 activity in human adherent monocytes: role of IL-1β and 15-HETE. Malar J 7:157

Prato M, Giribaldi G, Arese P (2009) Hemozoin triggers tumor necrosis factor alpha-mediated release of lysozyme by human adherent monocytes: new evidences on leukocyte degranulation in P. *falciparum* malaria. Asian Pac J Trop Med 2:35–40

Prato M, Gallo V, Giribaldi G et al (2010a) Role of the NF-κB transcription pathway in the haemozoin- and 15-HETE-mediated activation of matrix metalloproteinase-9 in human adherent monocytes. Cell Microbiol 12:1780–1791

Prato M, Gallo V, Valente E et al (2010b) Malarial pigment enhances Heat Shock Protein-27 in THP-1 cells: new perspectives for in vitro studies on monocyte apoptosis prevention. Asian Pac J Trop Med 3:934–938

Schwarzer E, Turrini F, Ulliers D et al (1992) Impairment of macrophage functions after ingestion of *Plasmodium falciparum*-infected erythrocytes or isolated malarial pigment. J Exp Med 176:1033–1041

Schwarzer E, Turrini F, Giribaldi G et al (1993) Phagocytosis of P. falciparum malarial pigment hemozoin by human monocytes inactivates monocyte protein kinase C. Biochim Biophys Acta 1181:51–54

Schwarzer E, Kuhn H, Valente E et al (2003) Malaria-parasitized erythrocytes and hemozoin nonenzymatically generate large amounts of hydroxy fatty acids that inhibit monocyte functions. Blood 101:722–728

Scorza T, Magez S, Brys L et al (1999) Hemozoin is a key factor in the induction of malaria-associated immunosuppression. Parasite Immunol 21:545–554

Serghides L, Patel SN, Ayi K et al (2009) Rosiglitazone modulates the innate immune response to Plasmodium falciparum infection and improves outcome in experimental cerebral malaria. J Infect Dis 199:1536–1545

Sharma G, Ottino P, Bazan N et al (2005) Epidermal and hepatocyte growth factors, but not keratinocyte growth factor, modulate protein kinase Calpha translocation to the plasma membrane through 15(S)-hydroxyeicosatetraenoic acid synthesis. J Biol Chem 280:7917–7924

Sherry BA, Alava G, Tracey KJ et al (1995) Malaria-specific metabolite hemozoin mediates the release of several potent endogenous pyrogens (TNF, MIP-1 alpha, and MIP-1 beta) in vitro, and altered thermoregulation in vivo. J Inflamm 45(2):85–96

Shio MT, Kassa FA, Bellemare MJ et al (2010) Innate inflammatory response to the malarial pigment hemozoin. Microbes Infect 12:889–899

Smith RJ, Sun FF, Iden SS et al (1981) An evaluation of the relationship between arachidonic acid lipoxygenation and human neutrophil degranulation. Clin Immunol Immunopathol 20(2): 157–169

Smith H, Wyke S, Tisdale M (2004) Role of protein kinase C and NF-kappaB in proteolysis-inducing factor-induced proteasome expression in C(2)C(12) myotubes. Br J Cancer 90: 1850–1857

Stenson WF, Parker CW (1980) Monohydroxyeicosatetraenoic acids (HETEs) induce degranulation of human neutrophils. J Immunol 124(5):2100–2104

Sue-A-Quan AK, Fialkow L, Vlahos CJ et al (1997) Inhibition of neutrophil oxidative burst and granule secretion by wortmannin: potential role of MAP kinase and renaturable kinases. J Cell Physiol 172:94–108

Urban BC, Todryk S (2006) Malaria pigment paralyzes dendritic cells. J Biol 5(2):4

van Phi L (1996) Transcriptional activation of the chicken lysozyme gene by NF-kappa Bp65 (RelA) and c-Rel, but not by NF-kappa Bp50. Biochem J 313(1):39–44

Verma I, Stevenson J (1997) IkappaB kinase: beginning, not the end. Proc Natl Acad Sci U S A 94:11758–11760

Wen Y, Gu J, Chakrabarti SK et al (2007) The role of 12/15-lipoxygenase in the expression of interleukin-6 and tumor necrosis factor-alpha in macrophages. Endocrinology 148(3): 1313–1322

Wen Y, Gu J, Vandenhoff GE et al (2008) Role of 12/15-lipoxygenase in the expression of MCP-1/CCL2 in mouse macrophages. Am J Physiol Heart Circ Physiol 294(4):H1933–H1938

Wyke S, Khal J, Tisdale M (2005) Signalling pathways in the induction of proteasome expression by proteolysis-inducing factor in murine myotubes. Cell Signal 17:67–75

Zhu J, Wu X, Goel S et al (2009) MAPK-activated protein kinase 2 differentially regulates plasmodium falciparum glycosylphosphatidylinositol-induced production of tumor necrosis factor-{alpha} and interleukin-12 in macrophages. J Biol Chem 284:15750–15761

Chapter 6
Human Lysozyme in Malaria Patients: Possible Role as Biomarker for Disease Severity

Mauro Prato, Manuela Polimeni, and Vivian Tullio

1 Introduction

Biomarkers are substances such as metabolites, enzymes, cytokines, and genetic markers that can be objectively measured in any biological fluids or specimens—including serum, plasma, urine, and cells—and used as indicators of physiological and pathological processes as well as responses to therapeutics. As such, biomarkers allow to support disease management and control by prosecuting different objectives: (1) to diagnose a disease; (2) to assess disease severity; (3) to provide a prognosis; (4) to provide some information relevant for the treatment and management of the disease; and (5) to identify the patients at risk for incurrence of the disease itself as well as long-term complications. Notably, the procedures required for biomarker evaluation must be simple, quantitative, and inexpensive therefore allowing for wide usage.

To date, biomarkers have been employed to diagnose and prognosticate the progress and outcome of several chronic and acute infectious, metabolic, and noncommunicable diseases, including cancer, diabetes, autoimmune diseases, and HIV/AIDS. However, the use of biomarkers in parasitic infectious diseases has been limited so far. As discussed in Chap. 1, the transition from mild malaria to the severe

---

M. Prato (✉)
Dipartimento di Neuroscienze, Università degli Studi di Torino, Torino, Italy

Dipartimento di Scienze della Sanità pubblica e Pediatriche, Università degli Studi di Torino, Torino, Italy
e-mail: mauro.prato@unito.it

M. Polimeni
Dipartimento di Oncologia, Università degli Studi di Torino, Torino, Italy

V. Tullio
Dipartimento di Scienze della Sanità pubblica e Pediatriche, Università degli Studi di Torino, Torino, Italy

M. Prato (ed.), *Human and Mosquito Lysozymes*,
DOI 10.1007/978-3-319-09432-8_6

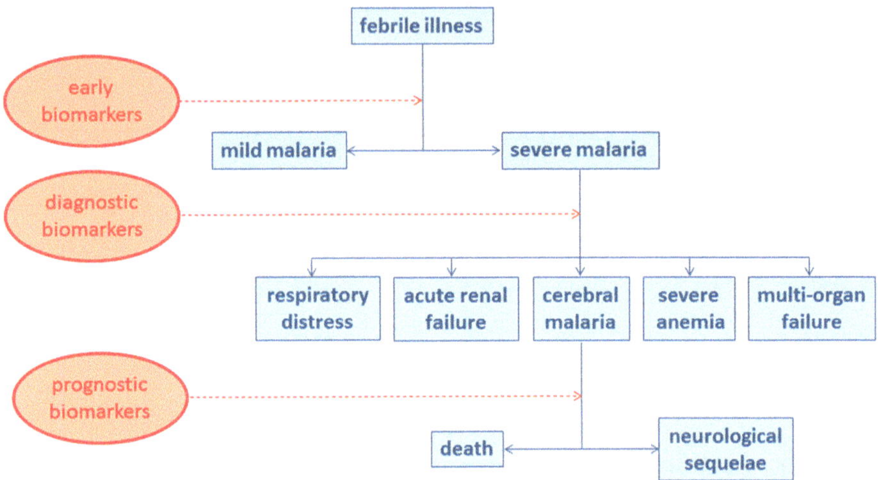

Fig. 6.1 Potential use of biomarkers in malaria management. The diagram briefly summarizes how early, diagnostic and prognostic biomarkers might be clinically employed during the various steps of the clinical course of malaria patients

forms of malaria can be sudden and requires immediate intervention. Therefore, the use of biomarkers to risk-stratify severe malaria patients would greatly enhance patient care and assist in appropriate management of health care resources (Lucchi et al. 2011). In addition, biomarkers able to identify asymptomatic patients, who might have levels of parasitemia undetectable by light microscopy or other conventional testing methods (see Chap. 2), could play a key role in monitoring the elimination of malaria reservoirs from endemic populations (Ogutu et al. 2010). The use of early, diagnostic, and prognostic biomarkers during the clinical course of malaria (see Fig. 6.1) might significantly improve the results obtained by malaria control and management programs.

Intriguingly, little yet consistent evidence is currently available from in vitro experiments (Prato et al. 2009; Polimeni et al. 2012, 2013) as well as from clinical studies (Mohamed et al. 1996; Mohammed et al. 2003) showing that increased levels of human lysozyme are associated with malaria severity. As it will be discussed in the next paragraphs, lysozyme can be detected in human fluids and specimens through several simple, quantitative, and inexpensive techniques. Therefore, a potential role for this molecule as an affordable biomarker of malaria severity has been recently proposed (Prato 2012).

Of course, it should be noted that the employment of biomarkers in the clinical practice requires a thorough validation of their utility. Collectively, five phases of biomarker development can be pinpointed (Pepe et al. 2001; Coca and Parikh 2008). The initial phase involves preclinical exploration during which potential biomarkers are identified. At this phase, markers capable of discriminating the severity of the disease are identified. During the second phase, new clinical assays measuring the potential biomarkers are established. In the third phase, retrospective longitudinal studies are carried out to test the new biomarkers for their utility, sensitivity, and specificity. The fourth phase involves a prospective screening phase which entails the use of the identified biomarkers to screen large populations. Finally, during the fifth phase the impact of the biomarkers on disease control and management is determined. In this context, the data supporting a role for human lysozyme as a marker of malaria severity have been scaled up to the third phase so far.

2 Methods to Measure Lysozyme Levels in Human Specimens

Since the discovery of lysozyme (Fleming 1922), several authors have proposed different methods for determining the activity of various preparations containing this enzyme. Most of these methods were based upon the clearing of dense suspensions of a susceptible organism without concern for accurate quantitative results. In 1946, the isolation of highly purified crystalline lysozyme by Alderton and Fevold suggested the possibility of a method for assaying lysozyme (Alderton and Fevold 1946). Since then, several quantitative methods to measure lysozyme levels in human fluids (serum, blood, urine) and specimens (tissues and cells) have been developed and standardized. In these days, several simple, cheap, and sensitive approaches routinely employed in research and diagnostic laboratories are available, including: (1) photometric assay; (2) radioimmunoassay (RIA); (3) immunohistochemistry (IHC); and (4) enzyme-linked immunoadsorbent assay (ELISA).

2.1 Photometric Assay

Lysozyme photometric assays have been developed after the methods proposed by Smolelis and Hartsell (1949) and Litwack (1955). These procedures provide for the rapid and accurate microbiological assay of fluid materials that show lytic activity considered to be due to lysozyme, including urine, serum, saliva, mucus, and tears (Osserman and Lawlor 1966; Seal et al. 1980).

The method is based on a comparison of light transmissions of standard crystalline lysozyme dilutions with the values for the substance being tested, after the addition of susceptible cells and incubation. A purified hen egg lysozyme is generally used as the standard. The natural substrate for lysozyme is the high molecular weight, insoluble peptidoglycan polymer that generally reinforces the bacterial cell wall. As such, this molecule—employed in the form of dried cells of UV-inactivated *Micrococcus lysodeikticus*—represents the most widely used substrate for measuring lysozyme lytic action on the linkages between *N*-acetylmuramic acid and *N*-acetylglucosamine in the cell wall. The cell suspension from the dried cells and the dilutions of crystalline lysozyme from a stock solution are prepared in Sorenson's phosphate buffer, pH 6.2. Rehydration is easily accomplished since lyophilized cells can be readily resuspended. Prior to the test the unknown should be checked for its activity to determine what dilutions are needed to give the same level of activity as the crystalline control. Before beginning the assay, a sufficient number of test tubes are matched so that all tubes will show the same light transmission value with distilled water as a reference. At measured intervals the lysozyme dilution is mixed with the cell suspension. The same procedure is used for the dilutions of the material being tested. All mixtures are made in duplicate.

After a 20-min incubation at room temperature, the light transmissions for the various mixtures are recorded and the concentration of the unknown is determined. With the transmission values for the crystalline lysozyme mixtures, a standard curve is prepared by plotting the transmission against concentration. A log scale is used on the abscissa of the standard curve. The transmission values for the unknown dilutions are located on the ordinate and projected to the standard curve. By projection to the abscissa from these points, the concentration of lysozyme in each dilution is determined. Multiplication by the dilution factor results in the concentration of lysozyme per mL of undiluted extract. The results obtained from this test are generally accurate and reliable.

2.2 Radioimmunoassay

The first studies to develop a RIA for human serum and urinary lysozyme date back to the late 1970s (Peeters et al. 1978; Thomas et al. 1981), as a consequence of the ongoing discussion concerning the diagnostic value of serum lysozyme determinations in gastrointestinal diseases, where differences in methodology were evoked as the cause of the observed discrepancies (Falchuk et al. 1975; Peeters et al. 1976).

Briefly, antiserum and labeled lysozyme are diluted in phosphate-buffered saline with 20 g/L of bovine serum albumin while standard dilutions are prepared from a stock lysozyme solution. Antiserum and labeled lysozyme are added to the standards and the diluted samples. The vials are incubated overnight at 4 °C, then the bound and free moieties are separated, vortex-mixed, and centrifuged. After aspirating the supernatant fluid, the radioactivity of the "free" fraction is counted. Samples of pooled sera at low, medium, and high lysozyme concentrations may serve as a quality control. Lysozyme results are interpolated from a curve, constructed by plotting the "free" count rate (in cpm) of standards versus the concentration of lysozyme.

2.3 Immunohistochemistry

IHC has become an increasingly popular and effective tool in research and diagnostic laboratories. The manifest success of this method is due to technological improvements, increased availability of purified antigens, antisera, and other reagents, and the accumulation of practical experience in a variety of applications. IHC combines the advantages of the high specificity and affinity of antibodies in recognizing tissue antigens with the high topographical resolution of light and electron microscopy. It can thus provide information which would be difficult or impossible to obtain by other techniques (Linnoila and Petrusz 1984).

In the respiratory system, for example, IHC investigations have demonstrated the presence of lysozyme in a number of different cell types, including serous cells of bronchial glands, neutrophils, and alveolar macrophages, pointing to the role of the enzyme as an antibacterial agent and as a diagnostic biomarker for pneumonia (Ishii et al. 2010). Specific staining was obtained with rabbit anti-rat lysozyme or anti-human urinary lysozyme antibodies but not with antibodies against egg white lysozyme.

2.4 *Enzyme-Linked Immunoadsorbent Assay*

Several ELISA kits for research or diagnostic purposes are commercially available to quantitatively measure lysozyme levels in human plasma, serum, urine, saliva, tears, mucus, and cell culture supernatants (Tiwari et al. 2012; Lu et al. 2014). This assay employs a quantitative competitive enzyme immunoassay technique that measures lysozyme rather quickly (~3 h). In general, an antibody specific for lysozyme is precoated onto a 96-well microplate with removable strips. Lysozyme in standards and samples can be competed with a biotinylated lysozyme sandwiched by the immobilized antibody and streptavidin-peroxidase conjugate. All unbound material is then washed away and a peroxidase enzyme substrate is added. The color development is stopped and the intensity of the color is measured. Lysozyme levels in the samples are finally calculated based on the generated standard curve.

3 Plasma Levels of Human Lysozyme in Malaria Patients

Although no retrospective longitudinal studies are available so far, the results from two small prospective studies enrolling healthy volunteers and malaria patients from Sudan have intriguingly shown enhanced lysozyme plasma levels in malaria patients, with a significant correlation between disease severity and lysozyme levels.

The first study (Mohamed et al. 1996) enrolled 14 patients with severe malaria cases in the pediatric casualty ward of Khartoum Teaching Hospital, Sudan, displaying convulsions, coma, and fever as clinical features with thick blood film positive for *P. falciparum*. Blood samples were collected before the administration of quinine, the standard treatment in such conditions. Lysozyme was estimated by RIA. Mean plasma lysozyme levels in patients were ~2200 µg/L, an amount almost double with respect to the corresponding levels in the controls. The differences between the levels of lysozyme in the patients and the controls were significant.

Interestingly, there was also a strong correlation between the levels of lysozyme and those of myeloperoxidase, another enzyme secreted by leucocytes (Dunn et al. 1968). The authors suggested that such a correlation may mean that both molecules were produced by the same cell type, or that the signal that activates myeloperoxidase production is also responsible for macrophage activation. It should be noted that the products of neutrophil degranulation are toxic to host tissues. Therefore, they may be responsible for the induction of severe malaria. However, both myeloperoxidase and lysozyme are produced also by monocytes/macrophages. Additionally, the study did not compare uncomplicated and severe malaria.

For this reason, the role of neutrophil activation in severe malaria patients was investigated by the same group in a second Sudanese study (Mohammed et al. 2003). Mohammed and colleagues enrolled three groups: (1) healthy individuals ($n=18$); (2) patients with uncomplicated malaria ($n=25$); and (3) patients with severe malaria ($n=22$). Moreover, they did not measure lysozyme and myeloperoxidase plasma levels only, but also human neutrophil lipocalin, a specific marker for neutrophil activation stored in neutrophil secondary granules (Xu et al. 1994). All malaria patients were *P. falciparum* infected, since patients infected with *P. vivax* or with mixed infections were excluded. Patients with mild, uncomplicated malaria received standard chloroquine, pyrimethamine/sulfadoxine, or quinine treatment.

Before starting the treatment, blood was collected from all individuals involved in the study. The plasma concentrations of lysozyme, lipocalin, and myeloperoxidase were measured by a double-antibody RIA. The severe malaria group showed significantly higher levels of all three proteins as compared with the mild malaria group or the group with no malaria. There were no significant differences between the mild malaria group and the group with no malaria. Interestingly, there was a highly significant correlation between the plasma levels of lipocalin and myeloperoxidase or lysozyme, respectively.

Notably, some studies have shown that the hemozoin-containing leucocyte count is a simple marker of disease severity in children with malaria (Amodu et al. 1997, 1998; Were et al. 2009). Consistently, as discussed in Chap. 5, the phagocytosis of hemozoin by human leucocytes has been associated with enhanced lysozyme secretion (Prato et al. 2009; Polimeni et al. 2012, 2013).

4 Conclusion

According to the results from two small prospective studies performed on Sudanese patients with malaria, human plasma lysozyme appears as a good candidate marker of disease severity. Therefore, future prospective as well as retrospective studies using larger cohorts of patients, possibly from different geographic areas, are certainly encouraged to assess the possible role of lysozyme as a biomarker for malaria.

References

Alderton G, Fevold HL (1946) Direct crystallization of lysozyme from egg white and some crystalline salts of lysozyme. J Biol Chem 164:1–5

Amodu OK, Adeyemo AA, Olumese PE et al (1997) Intraleucocyte malaria pigment in asymptomatic and uncomplicated malaria. East Afr Med J 74:714–716

Amodu OK, Adeyemo AA, Olumese PE et al (1998) Intraleucocytic malaria pigment and clinical severity of malaria in children. Trans R Soc Trop Med Hyg 92:54–56

Coca SG, Parikh CR (2008) Urinary biomarkers for acute kidney injury: perspectives on translation. Clin J Am Soc Nephrol 3:481–490

Dunn WB, Hardin JH, Spicer SS (1968) Ultrastructural localization of myeloperoxidase in human neutrophil and rabbit heterophil and eosinophil leucocytes. Blood 32:935–944

Falchuk KR, Perrotto JL, Isselbacher KJ (1975) Serum lysozyme in Crohn's disease and ulcerative colitis. N Engl J Med 292:395–397

Fleming A (1922) On a remarkable bacteriolytic element found in tissues and secretions. Proc R Soc Lond Ser B Biol Sci 93:306–317

Ishii H, Iwata A, Oka H et al (2010) Elevated serum levels of lysozyme in desquamative interstitial pneumonia. Intern Med 49:847–851

Linnoila I, Petrusz P (1984) Immunohistochemical techniques and their applications in the histopathology of the respiratory system. Environ Health Perspect 56:131–148

Litwack G (1955) Photometric determination of lysozyme activity. Proc Soc Exp Biol Med 89:401–403

Lu D, Li Q, Wu Z et al (2014) High-level recombinant human lysozyme expressed in milk of transgenic pigs can inhibit the growth of Escherichia coli in the duodenum and influence intestinal morphology of sucking pigs. PLoS One 9:e89130

Lucchi NW, Jain V, Wilson NO et al (2011) Potential serological biomarkers of cerebral malaria. Dis Markers 31:327–335

Mohamed AO, Elbashir MI, Ibrahim G et al (1996) Neutrophil leucocyte activation in severe malaria. Trans R Soc Trop Med Hyg 90:277

Mohammed AO, Elghazali G, Mohammed HB et al (2003) Human neutrophil lipocalin: a specific marker for neutrophil activation in severe Plasmodium falciparum malaria. Acta Trop 87: 279–285

Ogutu B, Tiono AB, Makanga M et al (2010) Treatment of asymptomatic carriers with artemether-lumefantrine: an opportunity to reduce the burden of malaria? Malar J 9:30

Osserman EF, Lawlor DP (1966) Serum and urinary lysozyme (muramidase) in monocytic and monomyelocytic leukemia. J Exp Med 124:921–952

Peeters TL, Vantrappen G, Geboes K (1976) Serum lysozyme levels in Crohn's disease and ulcerative colitis. Gut 17:300–305

Peeters TL, Depraetere YR, Vantrappen GR (1978) Radioimmunoassay for urinary lysozyme in human serum from leukemic patients. Clin Chem 24:2155–2157

Pepe MS, Etzioni R, Feng Z et al (2001) Phases of biomarker development for early detection of cancer. J Natl Cancer Inst 93:1054–1061

Polimeni M, Valente E, Aldieri E et al (2012) Haemozoin induces early cytokine-mediated lysozyme release from human monocytes through p38 MAPK- and NF-kappaB-dependent mechanisms. PLoS One 7(6):e39497

Polimeni M, Valente E, Aldieri E et al (2013) Role of 15-hydroxyeicosatetraenoic acid in hemozoin-induced lysozyme release from human adherent monocytes. Biofactors 39(3): 304–314

Prato M (2012) Human and mosquito lysozymes in malaria: old molecules for new approaches towards diagnosis, therapy and vector control. J Bacteriol Parasitol 3:e113

Prato M, Giribaldi G, Arese P (2009) Hemozoin triggers tumor necrosis factor alpha-mediated release of lysozyme by human adherent monocytes: new evidences on leucocyte degranulation in P. *falciparum* malaria. Asian Pac J Trop Med 2:35–40

Seal DV, Mackie IA, Coakes RL et al (1980) Quantitative tear lysozyme assay: a new technique for transporting specimens. Br J Ophthalmol 64:700–704

Smolelis AN, Hartsell SE (1949) The determination of lysozyme. J Bacteriol 58:731–736

Thomas MJ, Russo A, Craswell P et al (1981) Radioimmunoassay for serum and urinary lysozyme. Clin Chem 27:1223–1226

Tiwari S, Ali MJ, Balla MM et al (2012) Establishing human lacrimal gland cultures with secretory function. PLoS One 7:e29458

Were T, Davenport GC, Yamo EO et al (2009) Naturally acquired hemozoin by monocytes promotes suppression of RANTES in children with malarial anemia through an IL-10-dependent mechanism. Microbes Infect 11:811–819

Xu SY, Carlson M, Engström A et al (1994) Purification and characterization of a human neutrophil lipocalin (HNL) from the secondary granules of human neutrophils. Scand J Clin Lab Invest 54:365–376

Chapter 7
Beyond Lysozyme: Antimicrobial Peptides Against Malaria

Sarah D'Alessandro, Vivian Tullio, and Giuliana Giribaldi

1 Introduction

Antimicrobial peptides (AMPs) are components of innate immunity, the arm of the immune system in charge for the first defense against pathogens, not only in humans but also in plants, insects, and primitive multicellular organisms. AMPs are short amino acidic sequences with less than 100 residues with a secondary structure which can be used for their classification (Table 7.1) (Giuliani et al. 2007).

They have a broad spectrum of activity against many microorganisms like Gram positive and negative bacteria, fungi, and protozoa, but also viruses. Furthermore, antitumor activity for AMPs has also been reported (Hoskin and Ramamoorthy 2008).

AMPs have a rapid action (minutes to hours) but they are usually active in the micromolar range, at higher doses compared to other antibiotics.

Although the mechanisms of action of the majority of AMPs are not precisely defined, interference with membranes is recognized as the main activity. Figure 7.1 schematizes the most known hypotheses on the mode of action of AMPs on the membranes of microorganisms. The nonspecific activity on membranes gives to AMPs the advantage that they should be less prone to induce resistance in the target organisms, being their mechanism of action not connected to a specific target. However, this resistance-proof of AMPs has to be demonstrated. On the other side,

S. D'Alessandro (✉)
Dipartimento di Scienze Farmacologiche e Biomolecolari, Università degli Studi di Milano, Milan, Italy
e-mail: sarah.dalessandro@unimi.it

V. Tullio
Dipartimento di Oncologia, Università degli Studi di Torino, Torino, Italy

G. Giribaldi
Dipartimento di Scienze della Sanità Pubblica e Pediatriche, Università degli Studi di Torino, Torino, Italy

M. Prato (ed.), *Human and Mosquito Lysozymes*,
DOI 10.1007/978-3-319-09432-8_7

Table 7.1 Classification of AMPs

Structure	AMPs
Linear, no Cys	Cecropin A
Cys residues	Defensins
Rich in specific amino acids (proline, glycine, histidine, tryptophan)	PR39 (proline rich), Indolicidin (tryptophan rich)

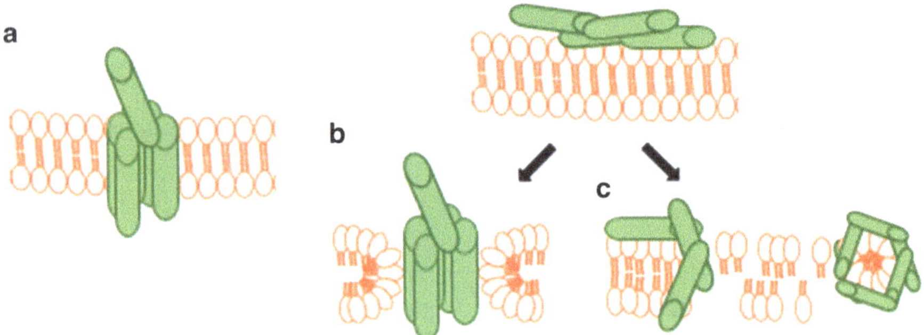

Fig. 7.1 The three model mechanisms of interaction between AMPs and biologic membranes. The image was modified from Chan et al. (2006). (**a**) Barrel/stave model. The AMPs form a pore in the membrane. (**b**) Torroidal pore. After massive AMP accumulation at the membrane surface, some AMPs acquire a transmembrane orientation and form pores, which have mixed composition (phospholipids and peptides). A curvature is induced in the membrane. (**c**) Carpet-like mechanism. The membrane surface, covered by AMPs, undergoes disruption

a disadvantage of the activity on cell membranes could be potential mammalian cell toxicity. This is true for some AMPs (e.g., gramicidin A, see paragraph 2.8), but not for others (e.g., some dermaseptin derivatives, see Sect. 2.6), which are specific for the membranes of microorganisms. In these cases, the difference in activity could be due to differences in lipid composition of membranes (cholesterol proportion or fluidity).

Beyond the activity at the membrane level, other intracellular targets such as protein or DNA synthesis have also been identified for some AMPs (Brogden 2005).

Due to their ability to penetrate cell membranes, AMPs have been proposed as vector for drug delivery (Splith and Neundorf 2011).

AMPs are difficult to be classified due to their huge diversity. The classifications can be based on different features, including amino acidic sequence (e.g., presence of cysteine residues, prevalence of particular amino acids, and presence of conserved sequences), membrane activity, secondary structure, and toxicity (Table 7.1).

2 Antimicrobial Peptides in Malaria

Some AMPs of different origin are known to affect *Plasmodium* development in different phases of the biological cycle, from asexual blood stages (cecropin, melittin, magainin, dermaseptin S4) to sexual stages in the mosquito, where AMPs can block ookinetes viability (VIDA 1-3, scorpine) or oocyst formation (VIDA 1-3) (Bell 2011). A recent work by Carter and colleagues investigated the effect of 33 AMPs on *Plasmodium* early sporogonic stages, verifying that they did not alter mosquitoes' fitness (Carter et al. 2013). Table 7.2 summarizes the antiplasmodial activity of some AMPs.

The secondary structure of AMPs has been used to predict the activity on different *Plasmodium* stages. For instance, Arrighi and colleagues designed new AMPs starting from natural or synthetic antimicrobial polypeptides and observed that peptides with no particular secondary structures (containing mainly random coils and turns) were more active on the sporogonic stages of *P. berghei* and *P. yoelii* (Arrighi et al. 2002).

Some antimalarial AMPs are hemolytic or toxic, whereas others specifically act on the membrane of infected red blood cells (RBCs) or directly on the membrane of the parasite and not on the membrane of uninfected RBCs. An example is given by dermaseptin S4, which is hemolytic and disrupts uninfected RBCs too. Development of more selective substitutes was necessary to decrease toxicity (Krugliak et al. 2000).

Table 7.2 Antimalarial activity of some representative AMPs

Activity	AMPs	Target	References
Inhibition of *Plasmodium* in vitro	Dermaseptin S4 (μM range)	Erythrocytic stages, especially trophozoites	Dagan et al. (2002), Ghosh et al. (1997), Krugliak et al. (2000)
	Vida 1-3	Ookinetes of *Pb* and *Py*	Arrighi et al. (2002)
	Scorpine	*Pb* ookinetes formation; asexual parasites	Carballar-Lejarazú et al. (2008), Conde et al. (2000)
	Cecropin, melittin, magainin e cecropin–melittin hybrids	Bloodstream forms	Boman et al. (1989), Gwadz et al. (1989), Wade et al. (1990)
Block malaria transmission in mosquitoes	Vida 1-3	Oocyst formation, *Pb* ookinetes in vitro, *Pb* and *Pf* sporogonic stages in mosquito	Arrighi et al. (2002), Carter et al. (2013)
	Defensin	Oocyst development	Shahabuddin et al. (1998)
	Melittin	*Pb* ookinetes in vitro, *Pb* and *Pf* sporogonic stages in mosquito	Carter et al. (2013)

2.1 Antimalarial AMPs Source

Antimalarial AMPs can be produced by mammalian hosts and mosquito vectors, as well as other organisms, which are not related to malaria (Table 7.3).

AMPs are part of the immune defense of mosquitoes, and *Plasmodium* infection can modulate AMPs expression in the *Anopheles* mosquito (Fig. 7.2). Vizioli and

Table 7.3 Sources of representative antimalarial AMPs

	AMPs	Origin
Human	Defensin	
Mosquito	Defensin	*A. gambiae*
	Gambicin	*A. gambiae*
	Cecropins	*A. gambiae*
Other organisms	Metalnikowin	*Palomena prasina*
	Scorpine	venom of *Pandinus imperator*
	Cecropin A	*Hyalophora cecropia*—Cecropia moth
	Magainin 2	Skin and stomach of *Xenopus laevis*

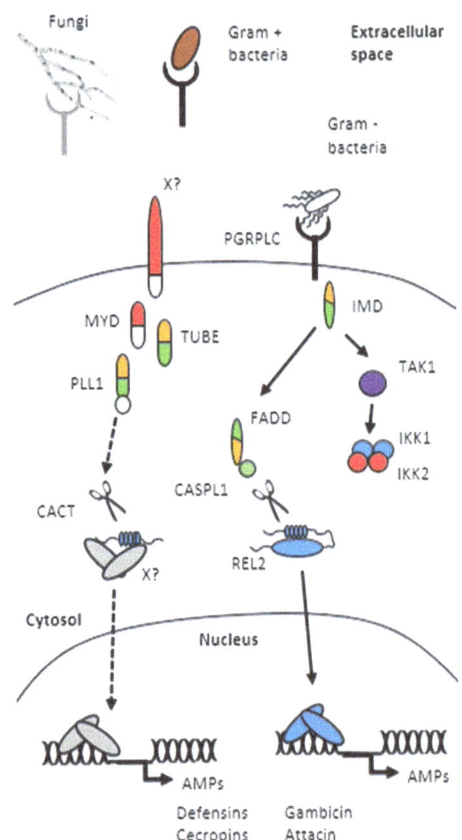

Fig. 7.2 The insect immune response to microorganisms. Common immune pathways in *Drosophila* and *Anopheles*. Proteins known in *Drosophila* with unknown ortholog in *Anopheles* are defined as "X?"

colleagues demonstrated that *Anopheles* mosquitoes fed upon mice infected with *P. berghei* expressed higher mRNA levels of cecropin A compared to mosquitoes fed with parasites unable to develop in the insect (Vizioli et al. 2000). Another example is described by Herrera-Ortiz and colleagues, who demonstrated that the mRNAs of attacin, cecropin, and gambacin were overexpressed in the midgut and abdominal tissue of mosquitoes fed with *P. berghei*-infected mouse blood (Herrera-Ortiz et al. 2011).

The majority of AMPs with antimalarial activity described by Carter and colleagues were derived from bee/wasp venoms (Carter et al. 2013).

Other examples of organisms producing AMPs with antimalarial activity are the scorpio *Pandinus imperator*, from which scorpine was isolated; the Cecropia moth which produces Cecropins; and *Xenopus laevis*, from which Magainin was extracted.

2.2 Defensins

Defensins represent the most important human AMPs as they are present at high concentrations (up to millimolar ranges) in epithelial and phagocytic cells. Their structure is characterized by a fold rich in beta-sheets and disulfide bonds between pairs of cysteines. The direct role of human defensins in malaria is not clear. Overexpression of a rat defensin (NP-1) was observed in a rat malaria model. Such enhancement was associated to protection of the young rats from lethal infection. That work supported a role for defensin in the immunity reaction to malaria infection (Pierrot et al. 2007). However, no direct studies on human defensins and malaria have been published.

Defensins are also part of the immune system of mosquitoes: their structure differs from that of human defensins, since it contains an alpha-helix linked to a beta-sheet. The role of mosquito defensins in malaria infection is better described compared to human defensins (Dixit et al. 2008; Hoffmann 1997; Meredith et al. 2008). Defensin expression, constitutive in mosquitoes midgut, is further induced by malaria infection (Richman et al. 1997; Vizioli et al. 2001b). The injection of defensin in *Aedes egypti* inhibited the development of *Plasmodium* sexual stages, resulting in oocyst abnormal development (Shahabuddin et al. 1998). The treatment of sporozoites with defensin decreased their viability.

However, a reverse genetic approach demonstrated that defensin is not necessary in *A. gambiae* (Blandin et al. 2002). The gene of defensin was disrupted in *A. gambiae* by treatment with dsRNA. This knockdown approach decreased the mosquito resistance to bacterial infections but did not alter the ookinete/oocyst formation or oocyst number after infection with *P. berghei*.

Table 7.4 Amino acid sequence of the major antimalarial AMPs discussed in this chapter

Name	Amino acid sequence
Scorpine	GWINEEKIQKKIDERMGNTVLGGMAKAIVHKMAKNEFQCM ANMDMLGNCEKHCQTSGEKGYCHGTKCKCGTPLSY
Cecropin A	KWKLFKKIEKVGQNIRDGIIKAGPAVAVVGQATQIAK
Melittin	GIGAVLKVLTTGLPALISWIKRKRQQ
CA(1-13) M(1-13)	KWKLFKKIEKVGQGIGAVLKVLTTGL
CA(1-8) M(1-18)	KWKLFKKIGIGAVLKVLTTGLPALIS
Magainin 2	GIGKFLHSAKKFGKAFVGEIMNS
Dermaseptin S4	ALWMTLLKKVLKAAAKAALNAVLVGANA
Gambicin	MVFAYAPTCARCKSIGARYCGYGYLNRKGVSCDGQTTINSCE DCKRKFGRCSDGFITECFL

CA cecropin A, *M* melittin

2.3 Scorpine

Scorpion venom is a rich source of peptides with different pharmacological activities. Interestingly, AMPs have been found in scorpion venom, and they may have different functions: the defense of scorpions from bacterial infection, the immobilization of their prey, or the synergistic activity with other venom toxins (Simard and Watt 1990).

In particular, scorpine (amino acid sequence in Table 7.4) is extracted from the venom of the scorpion *Pandinus imperator*. It was tested for the first time against *Plasmodium* due to its similarity, in the peptide sequence, to cecropins and defensins, already known for their antimalarial activity (Conde et al. 2000).

Scorpine decreased in a dose-dependent manner the fecundation of *P. berghei* parasites (measured as number of rosettes) and the formation of ookinetes (Conde et al. 2000). The inhibition of ookinetes formation in *P. berghei* was confirmed by Carballar-Lejarazù and colleagues, who also demonstrated the inhibition of asexual *P. falciparum* parasites in vitro (Carballar-Lejarazú et al. 2008). The authors used recombinant scorpine produced by transfected *A. gambiae* cells (cell line Sua 5.1). The plasmid for transfection was designed in order to make scorpine expressed under the control of the *A. gambiae* serpin promoter. They also created transgenic *Drosophila*, demonstrating that the expression of scorpine is not toxic to the insect. Such a paper was proposed as a proof of concept for the development of recombinant mosquitoes, an approach already proposed by Possani et al. (2002), as described below.

2.4 Cecropins, Melittin, and Cecropin–Melittin Hybrids

Cecropins are a group of insect-derived inducible antibiotic peptides from the giant silk moth *Hyalophora cecropia*. Cecropins A and B AMPs were fully characterized by Boman and colleagues, a work published by *Nature* and reproduced on

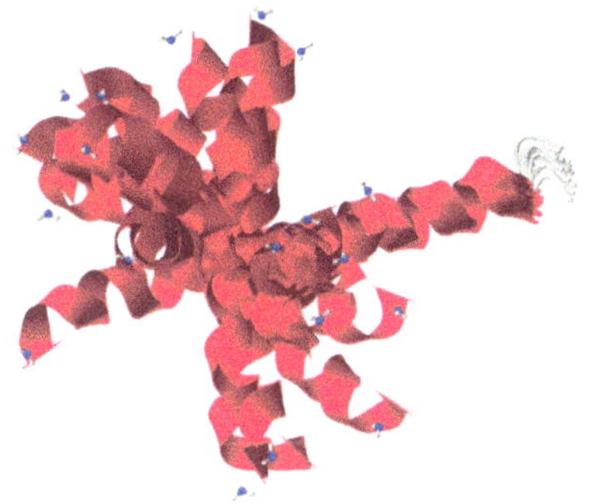

Fig. 7.3 Structure of cecropin. Image from the PFAM protein database (Punta et al. 2012) of the Wellcome Trust Sanger, Hinxton, UK (http://pfam.sanger.ac.uk/family/Cecropin)

The Journal of Immunology representing a pillar article in immunology (Steiner et al. 2009) (see Fig. 7.3 for cecropin structure). Cecropin B affected oocyst development in the *A. gambiae - P. cynomolgi* model (Gwadz et al. 1989). Some derivatives, namely Shiva-1, Shiva-2, and Shiva-3, were designed starting from the cecropin amino acidic sequence (Rodriguez et al. 1995; Yoshida et al. 2001). They inhibited the sexual stages of *P. berghei* as well as ookinete and sporozoite development in the mosquito model.

The structural conformation of melittin was described by Wade and colleagues as percentages of alpha-helixes, beta-sheet, and random coils (Wade et al. 1990).

Few years later the possibility of improving the antibacterial and antimalarial activities by creating hybrids between cecropin and melittin was explored (Boman et al. 1989). The properties of cecropins along with melittin and megainin to form ion channels in biologic membranes were studied in the 1990s (Wade et al. 1990). The amino acidic sequence of cecropin A, melittin, and two hybrids is reported in Table 7.4.

2.5 Magainin

Magainins were originally isolated from the skin of the African clawed frog *Xenopus laevis* (Zasloff 1987). Magainin (amino acidic sequence in Table 7.4) is active against different bacteria, such as *Escherichia coli*, *Streptococcus pyogenes*, and *Pseudomonas aeruginosa*, by forming pores in the membranes. Magainin affects the viability of others microorganisms, including *Saccharomyces cerevisiae* and *Plasmodium* spp (Gwadz et al. 1989). Some derivatives were developed. However, none of them were approved by FDA after clinical trials since they did not display increased activity compared to existing antibacterials or because they implicated toxicity issues. The structural conformation as percentages of alpha-helixes, beta-sheet, and random coils (see Fig. 7.4) was described by Wade and colleagues (Wade et al. 1990).

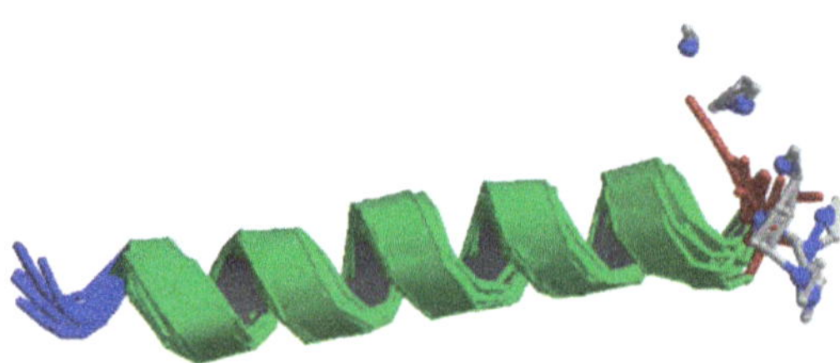

Fig. 7.4 NMR structure of magainin-2 in DPC micelles, ten structures. Picture from the Research Collaboratory for Structural Bioinformatics (RCSB) Protein Data Bank (Berman et al. 2000). Protein chains are colored from the N-terminal to the C-terminal using a rainbow (spectral) color gradient (http://www.rcsb.org/pdb/explore/explore.do?pdbId=2MAG)

2.6 Dermaseptins

Dermaseptins are a family of AMPs isolated from frogs of the *Phyllomedusa* genus with cytolytic activity against bacteria, protozoa, yeast, and filamentous fungi. Ghosh and colleagues compared hemolytic dermaseptin S4 (amino acidic sequence in Table 7.4) with nonhemolytic dermaseptin S3 for their physical properties (aggregation in solution and dissociation in membranes, binding to and interaction with RBCs) and for the effect on *P. falciparum* growth in vitro (Ghosh et al. 1997). Several derivatives were prepared starting from dermaseptin S4, with many showing a selective activity on the membrane of infected RBCs compared to the activity on the membranes of normal RBCs (Krugliak et al. 2000). The effects of dermaseptin S4 and its derivatives on malaria parasites were further investigated with respect to stage specificity (Dagan et al. 2002; Efron et al. 2002).

2.7 Gambicin

Gambicin (amino acidic sequence in Table 7.4) was first isolated from the conditioned medium of the *Anopheles gambiae* cell lines 4a-3A and 4a-3B (Vizioli et al. 2001a). The activity on different microorganism was tested and gambicin inhibited the growth of *Micrococcus luteus*, *E. coli* SBS363, and *Neurospora crassa*. Gambicin was also effective against *P. berghei* ookinetes. Moreover, as other AMPs, the expression of gambicin was enhanced by *Plasmodium* infection. In 2006, Dong and colleagues studied the immune response of *Anopheles gambiae* to the human *P. falciparum* or the murine *P. berghei* malaria parasites (ookinete stage) by DNA microarray analyses and RNAi gene silencing assays (Dong et al. 2006). The two species induced the expression of different genes and the authors confirmed the different ability to modulate the mosquito immune response to malaria.

2.8 Other Antimalarial AMPs

A possible classification of antimalarial AMPs is described by Bell (2011). Cationic, amphipathic "host-defense" peptides such as defensins and cecropins were treated in this chapter. Other membrane-active peptide antibiotics, such as gramicidin, have high activity on *Plasmodium* in the nanomolar range but they are also toxic for mammalian cells. Cyclosporine A, representative of the hydrophobic peptides class, was studied in all the *Plasmodium* stages and is active especially in the murine models. Thiopeptides, such as thiostrepton, have antimalarial activity but quite high IC_{50}. Some other naturally occurring or synthetic peptides have been shown to have antimalarial activity. The antiprotozoal activity of AMPs from amphibian origin was reviewed by Rivas and colleagues (Rivas et al. 2009).

3 Potential Application of AMPs in Malaria Research and Control

AMPs have been investigated as potential drugs against different *Plasmodium* stages and in particular against the erythrocytic phase, which is largely associated with the symptoms of the disease (Khadjavi et al. 2010). Recently, the interest of the research community and health authorities has moved toward elimination/eradication programs. To reach this ambitious goal, blocking transmission becomes an important step and AMPs could be reevaluated for their activities against the sexual stages, occurring throughout the mosquito vector.

The most described application for AMPs in malaria is mosquito and parasite engineering to reduce or interrupt malaria parasite transmission (Carter and Hurd 2010). Possani and colleagues proposed to insert the genetic code for bioactive peptides extracted from scorpion venom (scorpine mainly) into *Anopheles* mosquitoes to make them resistant to malaria infection (Possani et al. 2002). The authors started from evidence from the literature that *P. gallinaceum* ookinetes injected in *Drosophila melanogaster* were able to develop into sporozoites identical to those obtained in mosquitoes and, as expected, able to infect chickens. They designed a strategy involving *Drosophila* as an investigation tool to study AMPs toxicity against insects and *Plasmodium* development within the insect. However, the authors did not go beyond the design of this strategy and did not show results of the transgenic work, only referring to preliminary, encouraging results.

A big issue with these transgenesis approaches is represented by the ethical concern in releasing transgenic insects in the environment.

A different approach is to engineer those microorganisms living in mosquitos' midgut. In this case, the aim is to make the vector resistant to malaria parasites. *Metarhizium anisopliae* fungi were transfected with salivary gland and midgut peptide 1 (SM1), scorpine, or an antibody that agglutinates sporozoites. Mosquitoes were infected with this microorganism, leading to a reduction of sporozoites production by more than 50 %, with the best result, 98 % reduction, obtained with scorpine (Fang et al. 2011).

References

Arrighi RB, Nakamura C, Miyake J et al (2002) Design and activity of antimicrobial peptides against sporogonic-stage parasites causing murine malarias. Antimicrob Agents Chemother 46:2104–2110

Bell A (2011) Antimalarial peptides: the long and the short of it. Curr Pharm Des 17:2719–2731

Berman HM, Westbrook J, Feng Z et al (2000) The Protein Data Bank. Nucleic Acids Res 28:235–242

Blandin S, Moita LF, Köcher T et al (2002) Reverse genetics in the mosquito Anopheles gambiae: targeted disruption of the Defensin gene. EMBO Rep 3:852–856

Boman HG, Wade D, Boman IA et al (1989) Antibacterial and antimalarial properties of peptides that are cecropin-melittin hybrids. FEBS Lett 259:103–106

Brogden KA (2005) Antimicrobial peptides: pore formers or metabolic inhibitors in bacteria? Nat Rev Microbiol 3:238–250

Carballar-Lejarazú R, Rodríguez MH, De La Cruz Hernández-Hernández F et al (2008) Recombinant scorpine: a multifunctional antimicrobial peptide with activity against different pathogens. Cell Mol Life Sci 65:3081–3092

Carter V, Hurd H (2010) Choosing anti-Plasmodium molecules for genetically modifying mosquitoes: focus on peptides. Trends Parasitol 26:582–590

Carter V, Underhill A, Baber I et al (2013) Killer bee molecules: antimicrobial peptides as effector molecules to target sporogonic stages of Plasmodium. PLoS Pathog 9:e1003790

Chan DI, Prenner EJ, Vogel HJ (2006) Tryptophan- and arginine-rich antimicrobial peptides: structures and mechanisms of action. Biochim Biophys Acta 1758:1184–1202

Conde R, Zamudio FZ, Rodríguez MH et al (2000) Scorpine, an anti-malaria and anti-bacterial agent purified from scorpion venom. FEBS Lett 471:165–168

Dagan A, Efron L, Gaidukov L et al (2002) In vitro antiplasmodium effects of dermaseptin S4 derivatives. Antimicrob Agents Chemother 46:1059–1066

Dixit R, Sharma A, Patole MS et al (2008) Molecular and phylogenetic analysis of a novel salivary defensin cDNA from malaria vector Anopheles stephensi. Acta Trop 106:75–79

Dong Y, Aguilar R, Xi Z et al (2006) Anopheles gambiae immune responses to human and rodent Plasmodium parasite species. PLoS Pathog 2:e52

Efron L, Dagan A, Gaidukov L et al (2002) Direct interaction of dermaseptin S4 aminoheptanoyl derivative with intraerythrocytic malaria parasite leading to increased specific antiparasitic activity in culture. J Biol Chem 277:24067–24072

Fang W, Vega-Rodríguez J, Ghosh AK et al (2011) Development of transgenic fungi that kill human malaria parasites in mosquitoes. Science 331:1074–1077

Ghosh JK, Shaool D, Guillaud P et al (1997) Selective cytotoxicity of dermaseptin S3 toward intraerythrocytic Plasmodium falciparum and the underlying molecular basis. J Biol Chem 272:31609–31616

Giuliani A, Pirri G, Nicoletto S (2007) Antimicrobial peptides: an overview of a promising class of therapeutics. Central Eur J Biol 2:1–33

Gwadz RW, Kaslow D, Lee JY et al (1989) Effects of magainins and cecropins on the sporogonic development of malaria parasites in mosquitoes. Infect Immun 57:2628–2633

Herrera-Ortiz A, Martínez-Barnetche J, Smit N et al (2011) The effect of nitric oxide and hydrogen peroxide in the activation of the systemic immune response of Anopheles albimanus infected with Plasmodium berghei. Dev Comp Immunol 35:44–50

Hoffmann JA (1997) Immune responsiveness in vector insects. Proc Natl Acad Sci U S A 94:11152–11153

Hoskin DW, Ramamoorthy A (2008) Studies on anticancer activities of antimicrobial peptides. Biochim Biophys Acta 1778:357–375

Khadjavi A, Giribaldi G, Prato M (2010) From control to eradication of malaria: the end of being stuck in second gear? Asian Pac J Trop Med 3:412–420

Krugliak M, Feder R, Zolotarev VY et al (2000) Antimalarial activities of dermaseptin S4 derivatives. Antimicrob Agents Chemother 44:2442–2451

Meredith JM, Hurd H, Lehane MJ et al (2008) The malaria vector mosquito Anopheles gambiae expresses a suite of larval-specific defensin genes. Insect Mol Biol 17:103–112

Pierrot C, Adam E, Hot D et al (2007) Contribution of T cells and neutrophils in protection of young susceptible rats from fatal experimental malaria. J Immunol 178:1713–1722

Possani LD, Corona M, Zurita M et al (2002) From noxiustoxin to scorpine and possible transgenic mosquitoes resistant to malaria. Arch Med Res 33:398–404

Punta M, Coggill PC, Eberhardt RY et al (2012) The Pfam protein families database. Nucleic Acids Res 40:D290–D301

Richman AM, Dimopoulos G, Seeley D et al (1997) Plasmodium activates the innate immune response of Anopheles gambiae mosquitoes. EMBO J 16:6114–6119

Rivas L, Luque-Ortega JR, Andreu D (2009) Amphibian antimicrobial peptides and Protozoa: lessons from parasites. Biochim Biophys Acta 1788:1570–1581

Rodriguez MC, Zamudio F, Torres JA et al (1995) Effect of a cecropin-like synthetic peptide (Shiva-3) on the sporogonic development of Plasmodium berghei. Exp Parasitol 80:596–604

Shahabuddin M, Fields I, Bulet P et al (1998) Plasmodium gallinaceum: differential killing of some mosquito stages of the parasite by insect defensin. Exp Parasitol 89:103–112

Simard MJ, Watt DD (1990) Venoms and toxins. In: Polis GA (ed) The biology of scorpions. Stanford University Press, Stanford, pp 414–444

Splith K, Neundorf I (2011) Antimicrobial peptides with cell-penetrating peptide properties and vice versa. Eur Biophys J 40:387–397

Steiner H, Hultmark D, Engström A et al (2009) Sequence and specificity of two antibacterial proteins involved in insect immunity. Nature 292: 246-248 (1981). J Immunol 182: 6635–6637

Vizioli J, Bulet P, Charlet M et al (2000) Cloning and analysis of a cecropin gene from the malaria vector mosquito, Anopheles gambiae. Insect Mol Biol 9:75–84

Vizioli J, Bulet P, Hoffmann JA et al (2001a) Gambicin: a novel immune responsive antimicrobial peptide from the malaria vector Anopheles gambiae. Proc Natl Acad Sci U S A 98: 12630–12635

Vizioli J, Richman AM, Uttenweiler-Joseph S et al (2001b) The defensin peptide of the malaria vector mosquito Anopheles gambiae: antimicrobial activities and expression in adult mosquitoes. Insect Biochem Mol Biol 31:241–248

Wade D, Boman A, Wåhlin B et al (1990) All-D amino acid-containing channel-forming antibiotic peptides. Proc Natl Acad Sci U S A 87:4761–4765

Yoshida S, Ioka D, Matsuoka H et al (2001) Bacteria expressing single-chain immunotoxin inhibit malaria parasite development in mosquitoes. Mol Biochem Parasitol 113:89–96

Zasloff M (1987) Magainins, a class of antimicrobial peptides from Xenopus skin: isolation, characterization of two active forms, and partial cDNA sequence of a precursor. Proc Natl Acad Sci U S A 84:5449–5453

Index

M. Prato (ed.), *Human and Mosquito Lysozymes*,
DOI 10.1007/978-3-319-09432-8

Zeitfracht Medien GmbH
Ferdinand-Jühlke-Straße 7
99095 Erfurt, Deutschland
produktsicherheit@kolibri360.de